For the love of

scout

Promises to a small dog

*One dog's incredible journey
through life: from death row
to front row*

W9-CNC-741

Tracey Ison

Hubble & Hattie

The Hubble & Hattie imprint was launched in 2009 and is named in memory of two very special Westie sisters owned by Veloce's proprietors. Since the first book, many more have been added to the list, all with the same underlying objective: to be of real benefit to the species they cover, at the same time promoting compassion, understanding and respect between all animals (including human ones!) All Hubble & Hattie publications offer ethical, high quality content and presentation, plus great value for money.

Hubble&Hattie

More great books from Hubble and Hattie –

www.hubbleandhattie.com

First published in August 2017 by Veloce Publishing Limited, Veloce House, Parkway Farm Business Park, Middle Farm Way, Poundbury, Dorchester DT1 3AR, England. Fax 01305 250479 / e-mail info@hubbleandhattie.com / web www.hubbleandhattie.com.
ISBN: 978-1-845849-36-8 UPC: 6-36847-04936-2.

Contents

Foreword

by Dr SDJ Marston B Vet Med Vet MF Hom MRCVS

This is a story of bravery and heroics. A real page turner: you have to know how it ends.

Scout has hurdled through life leaping obstacles unseen and never-ending.

An indomitable spirit with the charisma of a superstar. Everyone is drawn to him and feels his supernova energy.

Tracey and Paul have gone far beyond commitment in their efforts for Scout, buoyed up and encouraged by the love and trust he gives back ...

Foreword

by Sarah Fisher

I have had the pleasure of sharing my life with several rescued dogs over the years, and *For the Love of Scout* resonated with me for many reasons. As someone who is passionate about the world of animal welfare, I am all too aware of the challenges faced by those who work so tirelessly to make the world a better place for dogs let down by some members of the human race.

This delightful book charts the story of a small, abandoned puppy, born with limited sight and suffering from mange and malnourishment. From a troubled past, and with an uncertain future, Scout was fortunate enough to secure a place on a transport run from a pound to East Midlands Dog Rescue. It was there he found his way into the arms, and hearts, of Tracey Ison and her husband, Paul.

As well as celebrating the joys of canine companionship, *For the Love of Scout* is an invaluable educational publication for anyone working or living with dogs. I love the diligence with which Scout's guardians began to lay important foundations on which his further education would be built: small, achievable, rewarding steps for both dog and humans alike.

If you ever feel disheartened or overwhelmed by the plight of unwanted animals, or if you are struggling with a dog in your care, pick up this inspiring book and immerse yourself in hope and possibilities. Remind yourself of all the good that exists, and smile as you read each promise Tracey made to the giving and forgiving Scout: promises that should be made to *every* dog.

Dedication

Dedicated to the memory of my dad, Alan. Keep sending the rainbows that light our way, and the white feathers that stick to Scout's nose and make him sneeze.

Introduction

When my husband and I first adopted Scout from a rescue centre in 2011, he was indeed a very small dog: a young puppy who had already endured so much in the first few months of his life. Back then, Scout had a vulnerability about him that melted us in an instant, and from the moment we first met this clumsy, flat-footed charmer, he stamped his pawprints well and truly onto our hearts.

On the day that we brought him home, I promised Scout that he would never have to suffer fear, loneliness, abandonment, and pain ever again.

Shortly after adopting Scout we were advised by one of my veterinary colleagues (who has never been backward in coming forward) that Scout may have a limited life expectancy, or, to use his own words "his hard wiring is likely to be all mixed up," meaning that the genetic problems that had manifested externally in Scout's colouring and sight issues could indicate that he wasn't quite right internally. This served only to strengthen our resolve to ensure that every day of Scout's life would be filled with love and kindness.

I have promised Scout that, no matter what his future holds, we will live for the here and now, and we will make the very best of every single day. I have promised Scout a life as normal as possible; that he won't be wrapped in cotton wool, and that we will meet every challenge head-on. I have promised Scout that his life matters, and that he deserves his place in this world.

These are the promises that I have made to our small dog.

1 Saving a life

"**W**hy have you never adopted a rescue dog?" was a question that I was frequently asked by family and friends. And why not, indeed? I have spent a lifetime knee-deep in dogs, cats, and assorted small, furry animals. As a qualified veterinary nurse for some twenty-five years and more, I have met countless waifs and strays of all different shapes and sizes, all crying out for help; all hoping for their chance to start a new life.

I also volunteered at an independent, UK-based dog rescue centre – East Midlands Dog Rescue (EMDR), run entirely by volunteers – where I met so many needy dogs in all sorts of desperate situations. I guess that if I allowed my heart to rule my head I would have a house-full by now.

I was content (or so I thought) with my existing little family of canine and feline companions.

Together with my husband, Paul, we shared our lives with two Whippets – Izzy and Misty – and a pair of senior cats, Jasmine and Daisy. Things seemed to be ticking along quite nicely, but little did we know that our lives were soon going to be turned upside down by the pitter-patter of tiny, clumsy paws ...

I was enjoying a sneaky cup of coffee at work one Monday morning, and found myself chatting to Anne, one of our reception team, who was filling me in on the new intake of dogs at EMDR. Our veterinary practice was the primary care provider for the dogs at the rescue kennels, and, like myself, my colleagues were always interested to hear about the comings and goings of this busy rescue centre.

> **"What cruel twist of fate had condemned this poor soul to a life spent in total darkness?"**

Sadly, like most rescue centres, East Midlands Dog Rescue was always full to bursting, with a never-ending list of dogs waiting to come in.

The rescue centre took in dogs from across the country, and from outside of the UK, too. Dogs snatched at the last minute from death row pounds by the aptly named 'pound pullers;' dogs who had suffered from severe abuse and neglect; ex-racing Greyhounds, whose careers had ended; ex-breeding dogs, and those dogs who were simply down on their luck and needed a safe haven. From pedigree dogs to crossbreeds, and elderly and infirm to young, effervescent puppies, there was a constant stream of needy dogs, all with a story to tell.

That weekend had been a busy one for the centre, according to Anne, and several new arrivals had checked into the kennels. One of these, Anne mentioned, just happened to be a Whippet-type puppy. My ears pricked up the instant I heard the word 'Whippet.' Whippets are, and always will be, the breed for me. Their effortless grace and elegance, combined with a sweet and loving nature, ticked every box on my list when looking for the perfect dog.

I asked Anne what she knew about the puppy, and she replied that, as yet, there were few details, but what she did know was that the pup had been earmarked for almost certain death at the stray pound where he had initially been taken, had cheated death by a whisker, and had secured a place on a rescue transport van that had travelled all the way from County Durham to Leicestershire. She added that it was widely thought the puppy was completely blind, and had been since birth.

My first thoughts were of huge sadness. What cruel twist of fate had condemned this poor soul to a life spent in total darkness?

I thought about the tragic plight of the puppy for the rest of the day. I hadn't even seen him, and knew very little about him, but something was niggling away at me. I kept thinking, could we squeeze in just one more? I arrived home that evening with thoughts of the puppy still going round and round in my head.

I decided to broach the matter with Paul, and dropped a little snippet about the puppy into a conversation in a roundabout way, testing the water, so to speak. Paul was deeply engrossed in reading a TV magazine when I casually mentioned that East Midlands Dog Rescue had taken in a little Whippet-type puppy, who had been saved from a pound, and what a terrible pity it was that one so young had found himself in such a desperate situation. I could sense that Paul was only half listening, as he continued to read. He nodded sympathetically, and agreed that it was a real shame, but no doubt the puppy would soon be snapped up.

I bided my time to allow Paul to digest the first piece of information, before pushing on and slipping into the conversation that it was also believed that this poor little mite was blind. Bullseye! *Now* I had Paul's full attention. Putting down his magazine Paul started firing off all sorts of questions about the puppy: how old was he; how did he come to find himself in rescue, and, lastly but most importantly, could we go and meet him?

I am not one for procrastination, and within minutes I had contacted Sandy, founder and driving force behind EMDR. Formed in the 1980s, the charity had grown, both in size and reputation, over the years, but was still very much a close-knit organisation run by like-minded people striving to do the very best for the dogs in their care. Sandy works tirelessly all day, every day, with a level of commitment, compassion and dedication that went above and beyond. She lives and breathes dog rescue, and, along with her partner, Rob, her 'right-hand woman' Lynne, and a handful of volunteers, she has successfully rehomed hundreds of dogs, relying solely on donations by the public to keep the centre afloat. I feel very privileged indeed to be able to call these people my friends.

Sandy was more than happy to fill me in on the puppy's background. Found straying on the streets of County Durham, he had been picked up and taken to a local stray pound. With no obvious means of identification, either by microchip or collar, he was held at the pound to complete his compulsory waiting period of seven days, to see if an owner stepped forward to claim him before being considered for rescue placement.

Concerns were raised about the puppy as soon as he entered the pound, as it was thought that he was both deaf and blind: he was also malnourished, suffered from mange, and had contracted kennel cough. Initial thoughts were that maybe it would be kinder to have him euthanised: after all, who would want to adopt a dog with so many problems? It seemed that the odds were well and truly stacked against this little pup, estimated to be only around four months old.

What he needed was a miracle, but miracles can sometimes be hard to come by ...

It transpired that the puppy's fate rested in the hands of two people: Sue and Lucille. Sue is the founder of GALA (Greyhound and Lurcher Aid), a charity involved in the rescue and rehoming of Sighthounds, and Lucille is the president of Lancky Dogs, a Lancashire-based group which, amongst other things, helped raise funds for GALA, and other small, independent rescues.

After due consideration, it was decided that Sue would try and assess the puppy's disabilities more closely, and she arranged to visit him at the pound. As the puppy had kennel cough and was potentially infectious to the other dogs in the pound, he had been moved to a cattery block to isolate him. Once Sue had confirmed his whereabouts, she let herself into his holding pen, and quietly sat down on the floor a short distance away from the little chap.

During his first few days at the pound it had become apparent that the puppy's vision was very poor, but a question mark hung over his ability to hear. If it was confirmed that he was both deaf *and* severely visually impaired, it seemed that euthanasia would be the kindest option. If he could hear, however, then maybe his life could be spared. Sue began to tap very lightly on the floor, to see if the puppy responded to the sound: a simple test that would mean the difference between life and death for him ...

The journey of hope

As if he knew that his life hung in the balance, right on cue the puppy lifted his head, got to his feet, and slowly made his way over to the source of the tapping. This simple act was all that was necessary to save his life.

His future secured, all the puppy needed now was a name, and Baby Buttons was what he was called during his stay at the rescue: 'Baby' because he was such a young dog, and 'Buttons' because he was described by one of his visitors as being 'as bright as a button.' Even in those early days this puppy was already melting hearts with his cheeky and affectionate nature.

Once Baby Buttons' compulsory waiting period had expired, the next job for Sue and Lucille was to find a suitable rescue placement for him. He still had to recuperate from his kennel cough, and also needed to have his vision fully assessed before being put up for adoption. Deciding where to place him was very straightforward, and a call was made to Sandy at East Midlands Dog Rescue, who immediately agreed to take Baby Buttons as soon as he was deemed fit enough to make the journey to the Midlands.

Due to the fund-raising efforts of members of the Lancky Dogs Club, adequate funds were available, not only to save Baby Buttons, but also the lives of twelve additional strays who remained unclaimed in the pound. Pooling resources again, Lancky Dogs and GALA drew up plans for an ambitious road trip to place all of the dogs in several different rescue centres spread across the UK; all to be achieved in just one day.

It was a cold and dark morning in February 2011 when the motley crew of thirteen waifs and strays embarked on what has now been dubbed 'the journey of hope' across the country to drop off each dog at the various rescue centres that had kindly offered space. Baby Buttons was joined on this epic journey by Wisp, Diamond, Snoop, Pearl, Rita, JJ, Major, Iris, Hobo, Mick, Mungo, and Tess: all Lurchers of varying shapes and sizes; all saved from a very uncertain future.

Lucille was part of the transport team that day, and completed the 600-mile round trip in fourteen hours, successfully delivering all of the dogs to their respective rescue centres. Baby Buttons and his new friend, a young Lurcher pup called Wisp, were left in the capable hands of Sandy at EMDR. All that he had to do now was wait ...

Meeting Baby Buttons

Paul and I first met Baby Buttons at East Midlands Dog Rescue one Sunday morning, about a week after he had been sprung from the pound. I had continued to enquire about how he was settling in, and Sandy was pleased with his progress: he was gaining weight, his skin condition was improving, and his kennel cough had almost resolved. Sandy said that he was a delightful little puppy who seemed to be coping remarkably well, despite the traumatic events that had occurred so early in his life. Sandy had also been observing him closely, and felt that, although Baby Buttons did have very poor vision, he

did not appear to be completely blind.

I am not sure quite how Paul was feeling at the prospect of meeting Baby Buttons, but I had a mixture of excitement, curiosity, and more than a few nervous butterflies fluttering around in my stomach as we walked through the main gates into the rescue centre.

Sandy took us straight through to the enclosed pen where Baby Buttons and his companion, Wisp, were being housed. Once inside, Paul and I knelt on the ground, and waited for Baby Buttons to make his entrance.

And he did just that! In a flurry of paws and waggy tails, out he ran with Wisp at his side.

My first thought was "What a strange-looking puppy!"

Baby Buttons gambolled towards us with the most peculiar, high-stepping gait, his front paws flicking outward at the wrists with each step he took. I can only imagine that this rather odd motion was allowing him to check the ground ahead for potential hazards and obstacles.

It would be dishonest of me if I said that Baby Buttons ran straight into our welcoming arms, but the truth is that I caught him as he barged past, and was instantly rewarded with a face full of soggy licks. I eventually managed to fend him off to get a better look at him.

Still very underweight, it was possible to see – and count – every single one of his ribs, and the spinous processes of his backbone looked like it might burst through the skin that was tightly stretched over it. His skin was very pink, and peppered with drying scabs, giving him a somewhat moth-eaten appearance. He also had an unpleasant-looking growth on the front of his chest.

In appearance, Baby Buttons looked very much like a Whippet; possessing the same athletic build combined with a long, graceful neck and muzzle, neatly folded back ears, and a deep chest supported by long, slim legs. On closer inspection, however, there were some very subtle differences. He had quite a broad skull, and disproportionately large, flat, front paws. He also sported a ruff of white fur around his neck, and a few sparse feathers on his tail.

The most striking thing about Baby Buttons was his coat, in both colour and pattern, with markings classified as red merle and white. The white parts of his coat were sparse, and exposed the sore and inflamed skin underneath; but the merle swatches of colour that interspersed these were a stunning mix of fawn, ginger, and cinnamon shades, all blending together in perfect harmony. His nose was bright pink with a brown splodge in the centre, making it look as if he had dipped his nose into a pot of melted chocolate. It was flanked on each side by an impressive set of white whiskers that quivered and twitched continuously as he moved around the pen.

The other outstanding characteristic about Baby Buttons were his eyes, or perhaps I should say his single visible eye. Appearing to be nothing more than an empty socket, closer inspection revealed a left eye deep within that had not properly formed. Baby Buttons' right eye was normal in shape and size, and a dazzling mix of blue shades that appeared ever-changing in both depth and intensity. Gazing into this eye, it was almost as if I was looking down a kaleidoscope, with its shifting patterns of vibrancy and hue.

My sound knowledge of genetics led me to conclude that Baby Buttons had been affected by a number of conditions, typically associated with dogs who carry the merle gene, that can be linked to both visual and auditory problems. The same genes responsible for Baby Buttons' distinctive coat pattern are also the reason why he was born with eye problems. The correct term for the deformity of his left eye is known as microphthalmia, and, as the name suggests, it means that the abnormally small left eye was totally obscured by the protective third eyelid.

The pupil of his right eye has spiky black projections radiating out into the blue iris – a so-called 'star burst pupil,' which means that the vision in this eye was likely to be very restricted indeed. The merle gene is also responsible for his aptly-named 'butterfly nose,' and pink colouration of his pads. It was heartbreaking to realise that the price Baby Buttons paid for his stunning coat and eye colour was his sight; a very cruel blow for this little puppy.

Baby Buttons nuzzled into my jacket, and, as I gently stroked the sparse ruff of fur around his neck, I felt a huge surge of emotion rushing through me: a warmth spreading from my heart, mixed with sadness that Baby Buttons had been condemned to live his life in almost total darkness. This little dog, whose life had hung in the balance, had beaten the odds; had survived out on the streets and in the pound, and had overcome serious illness. A strong and courageous heart beat within his painfully thin chest, and a determination to fight for his place in the world. Could we be the ones to guide him through life and to allow him to flourish? One look at Paul's face, wreathed in smiles, gave me my answer.

We left Baby Buttons that day capering in his pen with Wisp, looking carefree and happy as if the drama that had been such a major part of his early life was, to him, nothing more than a distant memory.

More than ever, now, I wanted us to be a part of his life; to be there to shape his future.

New friendships

Throughout the following week, Baby Buttons was never far from our thoughts. Paul and I agreed that we were well and truly smitten by him, and felt confident that we could offer him a good home.

We also felt a little apprehensive. We had never owned a dog with a disability before, so this would be a new experience. We also had to consider Izzy and Misty, and the impact that Baby Buttons would have on them: how would they cope, and would they accept him?

Izzy and Misty are a closely bonded pair. Izzy – a headstrong, self-assured dog who possesses strong leadership qualities – had assumed the role of guardian and protector to Misty from the first day that we brought her home as a ten-week-old puppy. Misty, on the other hand, is a quiet, biddable dog, happy to live in Izzy's shadow. Both are very sociable and well-balanced dogs, however, and we felt that they would make ideal companions for Baby Buttons.

We also spent the week reviewing the layout of our house and garden, removing what we felt could potentially be hazardous to Baby Buttons. This gave us the opportunity to do a spot of gardening – to cut back low-hanging branches, and prune the many shrubs and plants that filled our back garden. Everything was cut to above where we felt Baby Buttons' eye level was to try to reduce the risk of injuring himself when outside. I guess all of this confirmed what we already felt in our hearts: Baby Buttons was coming home.

The decider would be how Baby Buttons reacted to Izzy and Misty, and vice versa, so, with feelings of excitement and anticipation in equal measure, we took them to the rescue centre to meet their potential new brother.

Izzy and Misty were frequent visitors to East Midlands Dog Rescue, and so happy to greet Sandy and the other volunteers. After we had been there for a short time catching up on the latest news, Sandy collected Baby Buttons from his pen. Over he came with his characteristic bouncing stride, and greeted Paul and I like old friends. Izzy and Misty stood patiently to one side as he leapt around excitedly.

Izzy made the first move, as we had predicted, leaning over to give him a tentative sniff. The minute he felt her nose touching his fur, Baby Buttons whirled round, and as soon as he had locked on to Izzy's position, threw both front paws around her neck. Izzy was a little startled by this novel method of greeting, but stood still whilst Baby Buttons held her in a tight embrace. Once he had given her a thorough sniff, Baby Buttons released his grip, and then repeated the process on a very surprised-looking Misty, who also held her ground and accepted this very unusual way to say hello.

Once the introductions were over, we decided to take all three dogs for a short walk to see how they interacted whilst on their leads. This proved challenging, to say the least.

Izzy and Misty walk well on the lead, but Baby Buttons didn't seem to have a clue. He frolicked happily along the path, frequently falling off the kerb, twirling in sheer joy every few steps, and often ending up pointing in the opposite direction. He bumped into Izzy and Misty repeatedly, and at one point looked like nothing more than the ball in a pinball machine: bouncing off first Izzy and then colliding with Misty.

Our two stalwarts took his crazy behaviour in their stride with as much dignity as they could muster, and I lost count of the number of times we had to stop so that I could untangle Baby Buttons' lead from around his legs, receiving a succession of face licks each time.

We returned to the rescue kennels with all three dogs and walked them over to Sandy. Reaching her, I stopped once more to untangle the lead from around a very exuberant Baby Buttons, who promptly stumbled forward and landed head first in a bowl of water.

Sandy gently reminded us that we were not obliged to take Baby Buttons; there was still time to change our minds. With a look of steely determination on my face, I scooped him up in my arms, wiped his sodden face, and deposited a kiss on his pink nose. Change our minds? Perish the thought! No matter

what challenges lay ahead, we would face them together from now on.

Holding Baby Buttons, I softly whispered into his soggy ear that he now had a family for life. No more loneliness; no more fear; the first of many promises that we have made to our singularly unique little dog.

Homecoming

All that remained to be done was for Baby Buttons to pack his bags and say his goodbyes. His luggage did not amount to much – a dog-eared Mickey Mouse stuffed toy that he had taken a real shine to, and a smart new identification tag for his collar. He spent a little time with Wisp, who had been reserved and was just waiting to go to her new home, too; Sandy, and the other volunteers, also wanted to give him a parting cuddle, and wish him (and us) luck.

Eventually, we walked out of the gates of the rescue centre with Baby Buttons twirling happily at our side.

Our first road trip with Baby Buttons was a little fraught. We had settled Izzy and Misty on the back seat of our car, and had decided that it may be safer for Baby Buttons to travel upfront on my lap, where he obligingly curled up, snuggling into his Mickey Mouse toy. We were halfway home when Baby Buttons suddenly lifted his head, emitted a huge belch, and followed it with a stream of foul-smelling vomit. Pausing only to lick his lips, he promptly tucked his head into my jacket and went back to sleep, leaving me to hold down my feelings of nausea whilst sick trickled down my legs like warm, molten lava. The smell was overpowering, so the rest of our journey was completed with the windows wound fully down, and the air conditioning set to maximum.

> **❝ I'd love to say that his first trip over the threshold into his new home was a success, but the reality was that he walked head first into the side of a drainpipe ... ❞**

We were all very relieved to get home. Izzy and Misty quickly hopped out of the car and made their way up the path towards our back door. Gently, I roused Baby Buttons, lifted him off my lap and onto the ground. He stood for a few seconds sniffing the air around him, as if savouring the unfamiliar smells (or maybe clearing his nostrils of the unpleasant odour of vomit) before beginning to move. I'd love to say that his first trip over the threshold into his new home was a success, but the reality was that he walked head first into the side of a drainpipe. Undaunted, Baby Buttons simply shook himself off, and continued on his way until he caught up with his canine companions waiting patiently by the back door.

Once the door had been unlocked, Izzy and Misty went straight in, but Baby Buttons stopped at the doorstep with his front paws touching it. Carefully, he lifted each paw over the step and jumped inside. What he did next had us totally mesmerised.

Slowly, Baby Buttons began to make his way around the kitchen, keeping to the edges of the room wherever possible, using the same foot-flicking gait that we had observed at the rescue kennel. Each time he took a step he reached out with a front paw as if testing the floor in front for any obstructions; stretching his neck and

holding his head straight, whiskers constantly twitching, as he made his way slowly around the kitchen, along the hallway, and into the lounge.

We describe this method of Baby Buttons familiarising himself with his surroundings as 'mind-mapping,' as he seemed to be forming a mental picture of everything around him. We soon discovered that, amazingly, he had only to repeat this behaviour once or twice before he felt able to navigate with complete confidence. Paul and I were already beginning to realise just how much confidence Baby Buttons possessed as we watched him nose his way along the side of the sofa, before jumping assuredly onto it, curling up into a tight ball, his pink nose tucked under his tail. Giving a big, contented sigh, he fell fast asleep.

Our boy was home.

Settling in

One of the first things we did with Baby Buttons was change his name. This may appear a little unfair, as he was beginning to respond to this, but we decided that he needed a shorter, clearer-sounding name, in the hope that this would help with his training.

We drew up a short list of names, and discussed the merits of each one in turn, eventually deciding on Scout. My favourite book as a teenager was Harper Lee's classic *To Kill a Mockingbird*, and the spirit, courage and tenacity of the Scout in the book matched that of our Baby Buttons.

Over those next few weeks we really got to know Scout, and vice versa. It became evident that Scout was an optimist, who welcomed every new day as if it was going to be the best day ever! A spot of friendly competition developed between Paul and myself to see who would get downstairs first every morning to receive one of Scout's early morning greetings.

Scout would wake promptly to the sound of his name, eagerly climbing out of his bed, his tail already wagging. This was usually followed by a series of spins, play bows, and noisy barks before he launched himself at his greeter to smother them in licks. Scout took the greatest pleasure in the smallest of things. At breakfast and dinner times he would stand in front of his food bowl, and hop from one paw to the other in sheer excitement at the prospect of food. Sometimes, he would get a little carried away with this, and would hop straight into his food bowl, scattering his dried food everywhere. A trip into the garden was a perpetual delight to Scout, as he had skilfully mind-mapped its layout, and could negotiate his way around with ease.

Scout adored Izzy and Misty; they, in turn, had generously decided to accept him into their inner circle, and seemed to regard him very much as a cheeky younger brother who needed to be kept under close supervision at all times. They both appeared intrigued and, sometimes, quite baffled by his behaviour, and became adept at stepping politely out of the way when Scout decided to have a little run around the garden, taking the bumps and knocks that frequently came their way with good grace. They seemed to accept the fact that, if they chose to lie down in a sunny spot for a snooze, there was a very real chance they would be trampled on. Paul and I were also becoming accustomed to bruises from accidental collisions: we weren't as nimble on our toes as Izzy and Misty ...

During those early weeks, we took Scout to the vet several times to check his weight and monitor the progress of his skin condition. Our vet confirmed what we already suspected about Scout's vision: he was totally blind in his left eye, and had severely restricted sight in his right. We had been working on the assumption that Scout had a degree of vision in his right eye, because when he seemed to be looking at something or someone, he would swing his head to the right. Just how much he could see varied, depending on level of light and shade, with dull days allowing better vision, and bright sunlight limiting this severely.

Scout's skin continued to improve as we repeated monthly spot-on treatments to rid him of the mange mites that had caused him to lose so much fur. It was this same mite that had caused his skin to become so inflamed and sore. We had taken advice from one of Scout's vets regarding his food, and had chosen a complete diet rich in nutrients to help nourish and repair his damaged skin. The unpleasant-looking skin growth on Scout's chest, present when we first met him, began to reduce in size, and eventually disappeared completely.

We also applied a moisturising spray to Scout's skin regularly, which he grew to love very much. You could almost see the relief on his face every time the soothing liquid was sprayed over him, quickly followed by little grunts of contented pleasure as we gently massaged it into his skin.

Scout was steadily gaining weight and condition, and – slowly but surely – was beginning to blossom into a remarkably handsome dog.

2 Learning to walk

We began working on Scout's training from day one. It was clear that he was lacking in many skills – lead walking being the most obvious. He loved his walks, but they were always a bit of an experience.

We purchased a sturdy harness for Scout, to which could be attached a double-ended lead. One end of the lead we attached to a ring on the back of the harness, and the other end we fastened to a ring in front of his chest. This arrangement allowed us to help guide Scout in the right direction, which, initially, was forward rather than his preferred spin and reverse turn.

> **“ One minute, he was tripping merrily along at my side, and the next he was gone! ”**

We experimented with walking Scout by himself, and alongside Izzy and Misty, and soon realised that Scout seemed to cope much better when walking with company. Izzy and Misty took to walking either side of him, which helped to maintain his forward motion. We fitted the collars of Izzy and Misty with extra metal discs, so that they made more noise when they walked, and these helped Scout pinpoint their location. The knowledge that his canine companions were nearby seemed to boost Scout's confidence.

Even as experienced dog handlers, walking a visually-impaired dog was a steep learning curve for Paul and me. We quickly learned how to hazard-spot on Scout's behalf, with potholes, muddy puddles, tow bars and wheelie bins top of the list. My first (and last) experience of inadvertently walking Scout over a metal storm grate was memorable for all the wrong reasons. One minute, he was tripping merrily along at my side, and the next he was gone! I looked down to see that Scout had dropped straight down into the drain, all four of his legs having managed to slot neatly between the metal slats. Scout seemed quite unperturbed by the sudden shift in altitude, and tried to continue walking, his slender legs powering away, going nowhere. It took me a frantic few minutes to extricate him from his predicament. Scout took full advantage of the situation, and gave my face a thorough washing whilst he waited to be liberated.

We also began to learn that we could sometimes become so absorbed in watching every step that Scout took that we were oblivious to the goings-on around us, as was demonstrated one day when I was out walking all three dogs. We were heading out of our estate on a path next to a main road, meandering along in no real hurry, just enjoying the fresh air. A little way ahead I saw a bus turning right out of a side road. We passed the bus on the opposite pavement. It didn't really represent a hazard (or so I thought), so I kept my head down watching Scout as he pottered amiably by my side. A few seconds later, I felt something connect firmly with the back of my legs, and was suddenly and very unceremoniously propelled into a nearby hedge. Still gripping all three leads in one hand, I flailed about, trying to get back on my feet. Brushing twigs and leaves out of my face with my free hand, I was greeted by a sea of concerned canine faces: Izzy, Misty, and Scout seemingly a little disconcerted by my apparent sudden need to dive into nearby foliage.

I managed to extract myself from the twiggy embrace of the hedge, and, whilst brushing myself down, tried to figure out what had just happened. The bus that had turned onto the main road was now tootling off into the distance, and it dawned on me that the driver must not have made the turning properly, the front bumper of his bus therefore catching the back of my legs with just enough force to knock me off my feet.

Not one of my finest moments, it must be said.

On our more successful walks we made steady progress. We realised that Scout was very responsive to the slightest change in walking pace: if we slowed, he instinctively knew that there was something ahead of him, be it a kerb, parked car, or something obstructing his path. Scout would slow his pace to match ours, allowing us to guide him safely around the obstacle.

Feline frenzy

Some walks were a mini-adventure in themselves, and none more so than one encounter that we had with the local neighbourhood felines.

I woke one morning bright and early to see blue sky peeping through a tiny gap in our bedroom curtains, which always meant one thing to me: an early walk with the dogs.

On that particular morning, as it was my day off, I decided to do a little one-to-one walking with Scout, before taking all three out together to brush up on our loose lead walking skills.

Scout was his usual excitable self when I clipped a lead onto his harness, and managed at least three twirls before we got through our garden gate.

We started off well, with Scout trotting obligingly at my side, enabling me to maintain a loose lead whilst I chattered away to him about our plans for the day. We made our way around the estate at a steady pace, Scout stopping now and then to sniff some interesting scents. I was in no rush – more than happy to amble along.

We were on the home straight when I heard a noise coming from behind a hedge about fifty yards ahead of us. What started off as a low, guttural rumbling began to grow in intensity: the unmistakable sound of a posse of cats gearing up for a fight. Scout had obviously heard the noise, too, as his ears pricked up, and he started to hold his nose high up in the air, trying to pick up a scent. I carried on walking, assuming that the cats were fully occupied with their own goings-on, and wouldn't pay any attention to us at all. An error on my part, as it turned out.

We approached the hedge, just beyond which a group of four cats were having what can only be described as a stand-off, posturing and yowling at one another, seemingly oblivious to anyone or anything around them.

As we sauntered by, Scout suddenly decided that the noise was an open invitation to play, and, with no warning at all, suddenly bounded sideways into the fray, scattering cats in all directions. Adopting his best play bow, Scout proceeded to bark noisily. He knew there was something or someone there, and simply wanted to join in with what he thought was an exciting event. The cats, on the other hand, fuelled by adrenaline, and very much startled by the presence of an interloper, decided that this was not part of the game plan ... as one, four balls of supercharged feline fury hurled themselves at Scout.

I can honestly say I have never witnessed anything quite like it in all my days.

As assuredly as guided missiles the cats launched themselves on top of poor Scout, hissing, spitting, and emitting ear-piercing shrieks. Scout hadn't a chance, and was soon buried under a seething mass of fur.

I leapt into action straight away, literally plucking each cat off Scout's back, and depositing him or her as far away from him as I could. But, as fast as I removed one cat, another took their place. Every single one of them was very handy with their claws, of course, and not afraid to use them! A momentary lull in the melee gave me the chance I needed to scoop a very

bewildered Scout into my arms, and stuff him inside my coat. Scout began to paddle his legs, trying desperately to connect with solid ground, one of his back paws becoming wedged in my coat pocket in the process. This promptly and most unhelpfully tore, scattering keys, mobile phone, and dozens of scrunched up poo bags all over the path!

Still clinging onto Scout for dear life, I started to shuffle along the pavement, kicking my pocket contents in front of me in a last-ditch attempt to escape from the furious felines, who, by now, had decided that strength may just lie in unity, and sat, side-by-side, watching me disapprovingly. I picked up a bit of speed, getting the hang of the shuffle/kick manoeuvre, and eventually managed to get to the end of the road and around a corner.

Casting a quick look behind to check that the coast was clear, I gently lowered Scout to the ground and checked him over. Luckily, he hadn't sustained too many injuries. I could feel a few bumps along his back from where the cats' claws had dug in, but it could have been so much worse. In himself, Scout seemed to be more concerned about being picked up and carried than the fact that he had just been mauled by a gang of ill-tempered cats. He simply shook himself, wagged his tail, and began to walk off. I, on the other hand, had not fared so well. My hands were criss-crossed with angry-looking scratches, and my coat was badly shredded. I gathered my belongings from the path and we made our way home.

Paul was in the kitchen when I arrived, and received his usual greeting from Scout – a couple of twirls, followed by an obligatory cuddle. As he was cuddling Scout, Paul asked "Did you have a good walk, then?" My silence caused Paul to look up and notice my dishevelled appearance. "It's best you don't ask" I told him, as I opened a cupboard door to rummage for the first-aid box.

Another promise made to Scout, from that moment on: if we again heard the discordant tones of a caterwauling feline we would simply turn around and walk away ...

Training

As well as training Scout to walk in a straight line on a loose lead, we also taught him the 'wait' cue as we had with Izzy and Misty. Sighthounds can sometimes struggle with the 'sit' cue, as they can find the sit position quite awkward, due to their long legs; for this reason, the wait cue had always been what we used at kerbs for crossing roads, or whilst waiting to cross at pedestrian crossings.

Scout was no exception, and we encouraged him to wait by holding a tasty treat in front of his nose whilst he was in a standing position. Scout would stand perfectly still with his nose outstretched, sniffing the treat. Initially, we asked him to wait for just a second or two, pairing his action with the word 'wait.' If he obliged, he was praised and given the treat. Gradually, we increased the time we asked Scout to stand in the wait position, and continued to reward each successful wait.

As with all of Scout's training, we started off training him indoors with no distractions, moving then into the garden, and then out and about to quiet areas, building gradually to busier, noisier places. In this way, Scout was always set up to achieve his training goals – positive, reward-based training at its very best.

Training Scout was, and still is, a two-way process. As much as we were instrumental in teaching him how to walk safely on a lead, negotiate obstacles and avoid hazards, there were many things that Scout taught us in return.

Crossing roads at traffic lights with pedestrian crossings was one of these. Whenever we stopped at a set of traffic lights where there was a pedestrian crossing, we would give Scout the 'wait' cue when the lights were on red, holding him in this position until the lights changed to green. We noticed, however, that Scout would break his wait position early, before we were ready to move off, and it took us a while to realise that he was responding to the beeping noise that indicates to those with sight issues that it is safe to cross. Scout had worked out for himself that we tended to cross when the beeping noise was heard, and was just a little bit quicker off the mark than us. Clever boy, Scout!

We also taught Scout the word 'step,' both indoors and out.

This is a very useful training cue for a blind or partially-sighted dog. Steps can be a single step up or down – at a kerbside, say – or a full flight of stairs.

Once again, we had the help of tasty treats to achieve our goal.

Beginning in the house with no distractions, we positioned Scout at the bottom of the stairs, and gave him a tasty treat. One of us would then entice him forward slowly by holding another treat just in front of his nose. As he reached forward to the treat, Scout's front legs would bump against the bottom step, whereupon he lifted each leg in turn to place it on the step in front of him. Each time Scout stepped up, we said 'step,' and rewarded him with a treat. Eventually, Scout linked the cue word with the need to lift and place his paws.

We used this technique to teach Scout how to go up and down a flight of stairs, so that he was confident going both ways, and, as with every training technique we used with him, once he had mastered a training cue indoors, we moved outside and repeated the process, starting off in quiet areas, and graduating to busier ones, until Scout was confident at tackling all sorts of steps and stairs.

We also looked out for steps that were constructed from a range of different materials. As well as concrete steps, we tried wooden steps, solid metal steps, and rubber-coated steps. Scout was sometimes a little hesitant when confronted by new textures underfoot, but we took things very slowly when he appeared a little unsure, and were always quick to lavish him with praise once he felt confident enough to try them.

Teaching Scout the word 'jump' was also useful, as it enabled him to jump safely into and out of a car without the need to lift him. As this cue was a little more complex, we broke it down into a couple of stages.

Initially, we taught Scout to simply place his paws on the car seat, using treats as encouragement. Holding a treat in front of Scout's nose, we slowly moved it away from him to place it on the car seat, keeping it covered with one hand. At first, when we did this, Scout would stand still, and try to sniff out where the treat had gone, eventually placing both front paws on the seat as he tried to snuffle the treat from beneath our hand. Every time he did this we released the treat and praised him. We gradually progressed, moving the treat a little at a time along the car seat, until Scout not only placed his front paws on the seat but actually jumped up onto it to retrieve the treat. As soon as he did this, we began to pair his action with the cue 'jump,' and in next to no time Scout had this all figured out, jumping into and out of the car unaided.

As with any training technique, we kept each session short, and never set him up to fail. If he appeared to be tired or distracted, we simply stopped.

All in all, our promise to give Scout a lifetime of guidance and support was a promise well made.

Some aspects of Scout's training – as with the traffic lights – originated from Scout himself, and we were always quick to reward any positive behaviour that he exhibited.

Scout loved to play with toys, and we tried to buy him those that would appeal more to senses other than sight: noisy toys, balls with bells; soft toys filled with material that made crackling sounds when chewed.

Despite this, Scout's all-time favourite toy remained a plain and simple tennis ball, which he would come up with after a good rummage through his toy box. Carrying the ball around with him, periodically, he would drop it, and allow it to roll a short distance from him. Sniffing around on the floor until he either nudged the ball with his nose or brushed against it with one of his paws, Scout would pounce on it, pick it up, and repeat the process. All the time he was doing this his tail would be wagging and he would emit short, sharp barks of excitement.

To keep the game going, Scout learned to bring the ball to us so that we could gently roll it along the floor for him to rush after, retrieve, and bring back again. Scout never seemed to become frustrated if he couldn't find his ball because it rolled off a little too far, and his stoical patience was just one of the many things about Scout that we loved so much.

Freedom

One of the biggest dilemmas we faced with Scout as part of his general training was whether or not to allow him to run off-lead with Izzy and Misty. Scout loved to join in with his companions when they zoomed around the garden, and the skill with which he could safely negotiate his way around at speed was quite remarkable.

Whenever we let Izzy and Misty off their leads on a walk to allow them to let off steam, Scout would start to spin and bark wildly, as he grew increasingly frustrated at being restrained. Izzy was also becoming frustrated on his behalf, and took to running in circles around Scout, trying to encourage him to engage in a game of chase, which she loved so much. Scout's safety was always paramount, of course, and the prospect of letting our short-sighted pocket-rocket off-lead in an open space filled us with more than a little apprehension. On the other hand, we had made a promise to Scout to allow him to lead a full and normal life. What to do?

Scout had attended puppy training classes, in which we had managed to achieve a form of recall. Scout was always happy to come back when called, but often his sense of direction went rather awry, which was fine by us as his intentions, we knew, were always good. We were used to dodging side-to-side to try and intercept Scout as he sailed past us, heading in the direction that he *thought* our voices were coming from. I think we must have looked like goal keepers, as we stood, slightly crouched, poised to jump left or right at a moment's notice so that Scout could believe he had got it right. We discovered that sound does travel well, even in a slight breeze, and that Scout would always hone in on where he thought we were calling him, and make a beeline for us. This seemed like a good basis on which to focus a little more on Scout's off-lead skills.

Armed with a long training lead and a pocket full of treats, we began in earnest. We found ourselves a large, open field, free of hazards such as trees, ditches, and rabbit holes, and attached Scout to a special lightweight training lead, a good twenty feet (approximately six metres) in length. This allowed him the freedom to move a little further from us, but ensured that he would remain within a safe distance to begin with.

During these early stages of Scout's training, we kept Izzy and Misty on their leads (much to their disgust) to prevent Scout being distracted and becoming over-excited. We watched closely as Scout explored his surroundings, nosing about in the grass, and occasionally lifting his head to sample the different scents in the breeze that blew gently around him.

Every now and then we would call his name, and he would try and fix on where our voices were coming from, swinging his head to try and 'see' us with his right eye. Continuing to call him, putting excitement and urgency into our voices, Scout would make his way to where he gauged we were. We did have to keep moving to whatever location he was heading toward, but our aim was to avoid setting him up to fail. Every time Scout got close to us, we crouched low to the ground, opened our arms, and enveloped him in a big hug once he connected with one of us, which was the perfect reward, as far as he was concerned, showing his appreciation with a thorough face wash each time. We repeated the process time and time again, until it was firmly embedded in Scout's mind that returning when called was clearly the best thing ever!

The next step was to release our

hold on the training lead, which allowed Scout to wander a little further afield. We also began to let Misty off her lead again, safe in the knowledge that she was not likely to run amok with Scout, given the very sensible dog that she was.

Scout was more than happy to amble along at her side. Misty was never one for rushing about unless prompted by Izzy, who was still puzzled about why she wasn't allowed off *her* lead (letting Izzy off her lead with Scout would be the ultimate challenge). Capable of achieving lightning speed and breakneck turns, our fawn girl was an athlete through and through, and loved nothing more than demonstrating her running skills to anyone who cared to watch.

❝ Scatterbrained and hair trigger, Izzy was often the cause of Scout over-reaching himself, and ending up in a pickle ... ❞

Choosing a sunny day, with little or no breeze, we took all three dogs to the open field. After strolling around for a while, we let them off, one-by-one, starting with Misty. When it was Scout's turn, we just dropped the long lead rather then detaching it, as this gave us a feeling of security.

We let the two of them potter about for a while, before releasing Izzy from her lead.

Whoosh! Off she went, streaking past Misty and Scout, who must have felt a brief rush of wind as Izzy hurtled past.

Scout immediately took flight to run after her, picking up on the sound that the discs on her collar made as she headed off into the distance. Izzy continued to run at speed, gradually turning back towards us in a wide arc, whilst Scout continued in a straight line, heading further and further away.

Scout began to slow, then, and eventually stopped when he could no longer hear Izzy's collar discs. By this point. Izzy had returned to us. Scout lifted his nose in the air to try and pick up her scent.

We called his name, but our voices didn't seem to carry well, and we could see that Scout was beginning to get a little agitated. Hurriedly, I started to make my way over to him, but someone else jumped in ahead of me. Misty, who, until that point, had decided simply to watch the shenanigans, took off, and ran full speed toward Scout. Reaching him, she bestowed a reassuring ear lick; Scout instantly relaxed and returned the favour. Once Misty had established that

Scout was reassured, she began to gently nudge him in the right direction. As I continued to run towards them, both came trotting over at a relaxed pace, and I gathered them into a welcoming hug and rewarded them with a treat each.

From that moment on, Misty became our safety net whenever Scout was off his lead. She would watch out for him, quietly and unobtrusively, and the minute he showed any sign of straying a little too far out of his comfort zone, off she went to collect him. Izzy, on the other hand, never did learn how to extend a helping paw when it came to retrieving Scout. Scatterbrained and hair trigger, Izzy was often the cause of Scout over-reaching himself, and ending up in a pickle.

A promise to Scout to let him enjoy a life of freedom is one we can safely say has been well and truly fulfilled, I think.

Visit Hubble and Hattie on the web: www.hubbleandhattie.com
hubbleandhattie.blogspot.co.uk
• Details of all books • Special offers • Newsletter • New book news

24

3 Friends and survivors

Another promise we made to Scout was that he would be known and loved by a wide circle of friends.

Loving Scout was easy: he had such an easy-going nature, and had perfected the art of charming everyone who met him. Scout was adept at leaning sideways into people, refusing to budge until he had been stroked and petted. If someone knelt in order to get closer to him, Scout would wrap them in one of his special hugs with his paws, whilst smothering their faces with slobbery kisses, his tail wagging furiously.

Scout learnt how to modify his responses when greeting different types of people, exhibiting quieter, calmer behaviour when greeting the old and very young, and reserving his own special brand of giddy, clumsy exuberance for more robust members of the public.

Scout also had a talent for voice recognition. If he met an individual several times over, he began to recognise their voice. If certain people approached and called his name – even from a distance – Scout would instantly respond with a few spins, bows and barks, sometimes apparently barely able to contain his excitement at the prospect of meeting up with a friend.

Our friends and family have all taken Scout to their hearts, readily accepting him with all his quirks and funny ways.

In fact, Scout has helped us forge some of the closest friendships we have, which have grown and flourished over the years. Our common bond is the love that we have for our dogs, and together we have created a wealth of precious memories and shared experiences.

Friends – both two- and four-legged – come in all shapes and sizes, and I would like to introduce you to some of Scout's closest companions.

Firstly, there is Buddy, the Staffordshire Bull Terrier. Buddy is a survivor, like Scout, and a worthy ambassador for this much-maligned and misunderstood breed.

Buddy was found by his owner, Nicola, quite by chance. Locked in a tiny cage and abandoned, Buddy had almost certainly been left to die an agonising death, and, when Nicola found him, Buddy was close to death. He was severely malnourished, weighing in at a mere three kilograms: about a fifth of what his normal bodyweight should be. Buddy was suffering from demodectic mange: a skin mite that can cause extensive and, in some cases, irreversible skin damage. He had also been lying in his own urine and faeces, which had caused scalding to the pads of his paws. Sadly, it looked like Buddy had suffered months of neglect at the hands of previous owners.

Nicola didn't hesitate. She released Buddy from his metal prison, and drove nearly 90 miles to her vet, checking throughout the journey that Buddy was still alive. The vet felt that Buddy had little chance of survival, but Nicola was determined to give him a chance.

Buddy's journey back to health was a long and difficult one, but regular medicated baths, mite treatments, and a special diet slowly began to repair all of the damage done to Buddy's skin, although bad scarring as a result of the mites has left his skin vulnerable to infection.

Buddy has a very forgiving heart, and the scars that he bears on the outside are not replicated on the inside of this gentle and loving boy. Buddy now shares his life with Nicola and her husband, Jon, and was recently a guest of honour at their wedding.

Scout is particularly fond of Buddy. Buddy makes a lot of noise when he runs, meaning he is an ideal running companion for Scout, and because he also has very short, stubby legs, Scout can easily keep up with him. Being very robust in stature, Buddy is also impervious to the knocks and bumps so generously delivered by Scout.

❝ Neither Maria nor the vet could believe that Mila had survived her injuries as they were so severe ... ❞

Ex-racing rescue Greyhound Shadow, Saluki cross Bliss, and Mila (short for Milagro which in Spanish means miracle) a Galgo rescued from Spain, also feature largely in Scout's life.

Mila's story is a shocking one, that highlights the terrible plight of so many of these beautiful, elegant dogs. Galgos (also known as Spanish Greyhounds) are mainly used as hunting dogs in Spain, and are often kept in the most appalling conditions. At the end of each hunting

season, countless Galgos are either abandoned, or disposed of in the most inhumane ways.

Mila was one such Galgo, found abandoned in a field by a young Spanish couple who were out walking their dog. Horrified by what they found, they took Mila to a lady called Maria at a dog refuge called A New Day Spain.

Maria took in Mila and rushed her to a vet. It was thought that Mila had been struck around the head with a heavy implement in an attempt to kill her. Critically injured, Mila had then been callously dumped. Although the blows did not prove fatal, they had been inflicted with so much force that Mila had sustained a fractured skull. A deep wound in Mila's skull had become badly infected. Neither Maria nor the vet could believe that Mila had survived her injuries as they were so severe, but survive she did, and Mila went on to make a complete recovery.

Mila was transported to the UK, where a Greyhound-specific rescue – Birmingham Greyhound Protection – took Mila under its wing, and began the search for her forever home.

Our friend, Fay, had been following Mila's plight on social media – her story having attracted worldwide interest – and felt an overwhelming urge to step in and offer help. Being an experienced Sighthound owner for many years, Fay, and her husband, Darrin, believed they could offer the kind of home that Mila well and truly deserved.

Mila is now living the life that was almost denied her, along with her new canine companions, Bliss, the Saluki cross, and Shadow, the ex-racing Greyhound. Just like Buddy, Mila's ability to forgive and to trust again is humbling to see.

Two very special dogs who beat the odds, and who now have owners who will honour their promise to keep them safe for the rest of their lives. Both dogs bear physical scars that will never fade: lasting testimony to the cruelty they both endured.

Aptly-named, fuzzy hounds Gracie, Indy, and Izzy (another Izzy) are amongst Scout's favourite friends. This trio of small, rough-coated Lurchers, devoted companions to owners Neta and Jo, collectively give hours of entertainment with their clownish behaviour.

Gracie has recently been diagnosed with a rare eye condition that resulted in her becoming extremely sensitive to light. To remedy this, and, to relieve her discomfort, Gracie now wears a specially-made tinted visor: a bold fashion statement which Gracie manages to carry off to perfection.

Larry, the ex-racing Greyhound, is another one of Scout's closest companions. Adopted by our friend, Eluned, Larry has endured more than his fair share of problems, which – sadly for him – meant there was never going to be a queue of people waiting to adopt him. In fact, he had already been returned to the rescue that initially took him in because of his behaviour.

Larry had issues regarding his personal space. If his boundaries were crossed, Larry would respond with a warning growl, and, on more than one occasion, this was followed by a snap. Who knows why Larry had built such a defensive wall around himself? Maybe living a solitary life, deprived of human company and affection, had simply left him with no reason to trust people.

Larry felt secure in his own company, and saw no reason to change, but, with time, patience, and a lot of understanding, Eluned slowly broke through Larry's defences, sensing that, deep down, Larry was desperate to be loved, but just didn't know how to ask for this.

Larry is now a relaxed and happy boy who loves to be amongst company. The growls and snaps are long gone, and, to top it all, Larry recently became a successful blood donor, playing a major role in helping to save the lives of other canines. Never in her wildest dreams did Eluned think that a dog who hated to be touched unless it was on his own terms would go on to earn his stripes as a life-saver.

Susie, the Spaniel, is the most recent addition to Scout's extended family. Adopted in 2016 from East Midlands Dog Rescue by Jon and Nicola as a companion for Buddy, Susie is starting to discover just how it feels to be loved.

Lastly, but by no means, least, we have Cindy, our very own 'ginga ninja.' Owned and very much loved by Betty, Cindy, the feisty terrier was Scout's nemesis. You could put Scout and Cindy in a room full of dogs, and Scout would always find and accidentally trample on Cindy, who responded to being trodden on by giving Scout a firm telling off. Scout accepted these reprimands with good grace, and would quietly tiptoe away. Cindy is the matriarch of our group, who helps to keep everyone in line.

4 Showtime!

Friendship involves so many things, and those that Paul and I had were all about the creation of memories, exciting adventures, new experiences, and challenges that we faced together. Being amongst so many like-minded people meant that we tended to share common interests, one of which was our love of dog shows, where we met up with friends and their dogs most weekends throughout the summer.

Scout loved any kind of social event, and, as soon as we put him in his special show collar, he knew he was heading out on a day trip to meet his friends. He could hardly contain his excitement: barking, spinning, and play bowing until we put him in the car with Misty and Izzy and set off.

Attending dog shows, we have been to some beautiful locations, and enjoyed some amazing events, festivals, fetes, and country fairs. The shows vary greatly, too: small events held in a tiny corner of a field at a village fete, right up to those run by professional groups, which are always hugely popular, and attract a large following.

We are great supporters of a lovely couple called Brian and Sue, who run many very successful dog shows under the Heart of England Dog Show banner. Brian and Sue organise a series of shows over the summer (and some indoor events during the winter months), and, at each show, a number of 'golden tickets' are awarded to those dogs who won certain classes that qualified them to attend a championship event at the end of the season, culminating in one winning the much-coveted 'Champion of Champions' title.

> **" My stomach was in knots when we took our places to be judged; Scout, on the other hand, was calmness personified ... "**

During the summer of 2014, Scout had been lucky enough to win a golden ticket in one of the crossbreed classes at a local show, qualifying him to enter the championship show, due to be held later that year, in October.

On the day of the championship show, we were full of excitement. Sporting a snazzy new show collar, Scout had been groomed until his coat shone; his nails has been neatly trimmed, and he had even had his teeth brushed for the occasion. Arriving at the indoor venue in good time, we settled ourselves with our friends to enjoy the show.

Eventually, Scout's class was announced. This class was for all dogs who had won a golden ticket in the crossbreed classes held at various shows during the summer months. Scout proudly trotted around the showring with his characteristic high-stepping gait: nose held high, scenting the air; tail swishing rhythmically, side to side.

The judge looked at each dog in turn, and, during his turn, Scout unleashed a full-on charm offensive, leaning into the judge's leg for a belly rub. The judge, enamoured by this behaviour, knelt to give Scout a proper greeting, and was instantly smothered in warm, wet licks. This may have been what clinched it, as Scout was announced winner of the overall crossbreed champion class, which placed him in the grand final, with the other class winners that day.

My stomach was in knots when we took our places to be judged; Scout, on the other hand, was calmness personified. He had enjoyed a wonderful day with his friends, and was making the most of being the centre of attention: win or lose, it mattered not one jot to him.

The judge took her time, and appraised each dog again before declaring Scout overall Champion of Champions!

All I could hear was the sound of people clapping and cheering. I couldn't see for the tears blurring my vision. Kneeling to give Scout a big hug, he promptly returned the favour by licking away my tears. I could have simply burst with pride.

We were called to the front of the showring, where Scout was awarded a stunning rosette, beautifully crafted in red, white and blue, with the words *Champion of Champions* emblazoned on one of its ribboned tails.

The realisation that those at the show truly appreciated Scout's beauty – both internally and externally – made my heart glow.

Visit Hubble and Hattie on the web: www.hubbleandhattie.com
hubbleandhattie.blogspot.co.uk
• Details of all books • Special offers • Newsletter • New book news

5 Pup Aid

Throughout 2014 we continued to embark on new adventures, and enjoyed many new experiences as a result, and one of the most memorable was the day we spent at the annual Pup Aid event on Primrose Hill in London.

Pup Aid was founded in 2010 by TV vet Marc Abraham, and helped raise awareness about the dark and sinister world of puppy farming, as well as promoting responsible dog breeding, and encouraging people to consider adopting a rescue dog. This event seemed to tick all of our boxes: a day out with friends, both four- and two-legged; a chance to meet some of the celebrities invited to judge the fun dog show, and, as ever, the opportunity to support animal welfare causes very close to our hearts. That this meant a longish car journey into the heart of London seemed a trifling matter ... our minds were made up; we were going!

So it was that, with a great deal of excitement, we made our way to London with Scout, on September 6. Deciding that Izzy and Misty would probably prefer a 'sofa day' at home rather than a long day out in what we thought would be a crowded environment, we left them comfortably ensconced on the sofa with a new chew apiece, and the promise of a visit from our local dog walker later in the day. They seemed more than happy with that decision.

It was slightly chilly, and cloudy, on Primrose Hill when we arrived that morning, but this didn't dampen our spirits as we made our way toward the brightly-coloured stalls encircling the showground. Some clever forward planning involved arranging to meet up with friends, Jo and Neta, who had travelled in style to the event by train, with their dogs, Izzy and Gracie. Not wanting to miss a single moment (we especially wanted to see the inspirational duo Owen and Haatchi, who had been invited along to open the event), we were there in plenty of time.

Owen was born with a rare genetic disorder that left him confined to a wheelchair. Haatchi, an Anatolian shepherd dog, had been found, left for dead, on a railway line, having suffered horrific injuries, which included the loss of one of his hind legs. Owen and Haatchi were brought together, and, from that moment, a very beautiful and enduring relationship was born. The bond that existed between this little boy and his dog touched the heart in so many ways, and the exceptional courage shown by both was testimony to their spirits and determination.

We set off to wander around the many stalls there – paradise for dog lovers like us! Scout, Izzy and Gracie soon busied themselves with sampling many of the dog treats on offer as we worked our way around.

Registration for the fun dog show was top of our agenda, and we were more than a little intrigued by some of the class descriptions. As well as the usual classes – Golden Oldies, Prettiest Bitch and Handsome Dog – some slightly more unusual-sounding classes also featured, such as Dog Who Looks Most Like a Celebrity, and Most Stylish Pooch in London. As we had armed ourselves with a variety of fancy dress, we signed up for these classes, fully entering into the spirit of the day.

The dog show kicked off mid-morning, with some of the early classes followed by a display from the Hearing Dogs charity, which provided such an admirable service. The amazing dogs in their care are trained to alert their deaf owners to important sounds and danger signals, both inside and outside the home, and is yet another reminder of just how much our canine companions help us.

After the Hearing Dogs display, it was time for Scout to make his debut in the showring for the Dog Who Looks Most Like a Celebrity class. Sporting a very fluffy mane, and four fluffy feet, we introduced Scout to the judges as Clarence, the cross-eyed lion. He was joined in the showring by Barney the Boxer as politician Boris Johnson, Ollie the Chihuahua as Liberace, and Chino the Tibetan Terrier as Cruella de Ville, to name but a few. No placing for Scout, but Gracie was awarded a very respectable fifth for her portrayal of former Spice Girl Ginger Spice.

The next class was the Best Rescue event. Several judges had been gathered to judge this busy class, and they made their way slowly around the ring, stopping to ask each owner about their dog. Scout was once more in his element as each judge stopped to give him a cuddle, receiving his full repertoire of face-licks, cuddles, and tail-wags in return.

Celebrity judge Peter Egan – champion of many animal welfare campaigns – seemed especially affected by Scout's story, as were the other judges who spoke to us that day. Peter had tears in this eyes as Scout greeted him by putting both paws on Peter's chest, leaning in for a cuddle. As he left us, Peter looked back, patted his chest with his hand, and mouthed "He has stolen my heart."

❝ ... we waited with bated breath for the result. By this point, Scout had decided that it was high time he took a nap, and had stretched out on the grass for a quick snooze. ❞

Patiently, we waited whilst judging was completed, the judges huddled together in the centre of the ring. More than once I heard mention of Scout's name, and I began to hope that maybe he would be placed in this very special class – what an achievement that would be! Imagine my great surprise and delight, then, when it was announced that the judges' unanimous decision was that Scout was the winner of the rescue class! On legs that had turned to jelly, and a little unsteady on my feet, we walked over to receive Scout's red rosette. Scout marched confidently into the centre of the showring, high-stepping all the way. The crowds gathered around the ring cheered and clapped Scout, each holding a little bit of our boy in their hearts that day. Pleased as punch, we made our way back to Paul, Neta and Jo, who were smiling, ear-to-ear. It was only then that the realisation dawned on me that Scout would be in the grand final of the Pup Aid dog show 2014 – how fantastic was that?!

As we continued to enjoy our day out with our friends, we were touched by the number of people who came over to say hello to Scout, and to ask about his story. Scout, as ever, continued to play to the crowd, loving every cuddle, stroke and treat that came his way.

Eventually, the time came for Scout to join his fellow class winners in the line-up for Best in Show, and, proudly taking our place, we waited for judging to commence. As with the rescue class, there were several judges, who, once again, stopped for a brief chat with each dog owner before making their all-important decision.

The first award presented in the Grand Final was Reserve Best in Show, and we waited with bated breath for the result. By this point, Scout had decided that it was high time he took a nap, and had stretched out on the grass for a quick snooze. I sat next to him and gave him a gentle tummy rub. I was so engrossed in what I was doing that I almost missed Scout's name being called – he had won Reserve Best in Show! To say I was proud would have been a massive understatement, and, gently nudging Scout awake, we once more joined the judges

in the centre of the ring to receive our gorgeous (and huge) rosette, claps and cheers ringing in our ears. The very handsome and much-deserving Boxer, Barney, was overall winner, and a popular one, at that.

It was the perfect end to a perfect day.

The invaluable work carried out by Pup Aid's army of supporters is tremendous, as they tirelessly campaign to bring an end to the cruel practice of puppy farming. One of the events at Pup Aid 2014 was a parade of ex-breeding dogs, all rescued from dreadful puppy farms. This small collection of broken dogs symbolised the suffering and hardship that thousands of others like them endure on a daily basis for years. The sadness etched on the faces of the dogs, as they slowly plodded around the ring, told stories of lives lived in isolation, squalor, and deprivation. They were few in number that day, perhaps because so many ex-puppy farm dogs would simply not have been able to cope with the noise and crowds. These dogs had suffered – and continued to do so – from the legacy of multiple health and psychological problems that could and should have been attended to earlier in their lives. Because they do not have their own voice, it is down to us to speak out for them, by supporting organisations like Pup Aid, and help stamp out this vile trade forever.

During my career as a veterinary nurse, I have known about and dealt with the consequences and casualties of puppy farms, and have first-hand experience of what

(continued on page 41)

Scout in the Lancky Dogs transport van.

Scout sleeping peacefully in Paul's arms shortly after we adopted him.

Scout at around six months old: very thin but improving daily.

Izzy. (Courtesy Bee Jackson www.happypetsphotography.com)

Scout running fast and free.
(Courtesy Sue Vought www.svphotography.co.uk)

Scout, Izzy, and Misty 'helping' me at work.
(Courtesy www.juliebournerphotography.com)

Larry.
(Courtesy Jo Green)

Scout, Izzy, and
Misty playing fetch.
(Courtesy Bee Jackson
www.happypetsphotography.com)

Misty.
(Courtesy Bee Jackson
www.happypetsphotography.com)

Scout in his
walking harness.

Neta with Gracie,
Indy and Izzy.
(Courtesy Jo Green)

Nicola, Jon, Susie,
and Buddy.
(Courtesy Jo Green)

Best friends!
(Courtesy Sue Vought www.svphotography.co.uk)

Fay, Darrin, Mila,
Shadow, and Bliss.
(Courtesy Jo Green)

Sparky, the problem whippet.
(Courtesy Bee Jackson www.happypetsphotography.com)

Scout demonstrating his ability to charm the judges at a local dog show.
(Courtesy Jo Green)

Cindy.
(Courtesy Jo Green)

Scout with his rosettes at Pup Aid 2014.
(Courtesy Jo Green)

Flying Whippets Sparky, Izzy, and Scout.

Scout with some of his most memorable awards.

Scout and I with Fay and Hero, after receiving awards in a special needs class.

Scout and his other best friend, Mickey Mouse.
(Courtesy Jo Green)

Scout with his winner's rosette for the RSPCA's Ruffs Perfectly Imperfect class.

Team whippet: Misty, Scout, Sparky, and Izzy.

Fabulous beach fun with friends!
(Courtesy Jo Green)

Scout and Sparky having great fun on the beach at Sea Palling.(Courtesy Jo Green)

Scout, Izzy, Misty, and Paul on Scout's first beach holiday to Wales.

Sweet treat for Scout.
(Courtesy Jo Green)

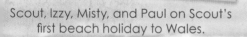

Run, Scout, run!
(Courtesy Jo Green)

Halloween fancy dress at Holkham Hall, Norfolk. (Courtesy Jo Green)

happens 'behind closed doors.' These sickening stories are beyond the scope of this book, but I would like to share with you the touching story of Molly...

Molly the Bull Mastiff had the misfortune to be born on a Welsh puppy farm in 2005, where, for the first few years of her life, she was kept in an outdoor pen, and used as a breeding bitch. It's likely that Molly was bred from repeatedly until she approached middle-to-old-age, when, as is usual, litter sizes reduce, and profit margins decrease, Molly was signed over to a local rescue.

As is the case with many rescues, Molly was subsequently booked in to be neutered to end her breeding career once and for all. Sadly, during her neutering operation, a cancerous growth was found, and Molly was given only a short time to live. Molly's rescue centre was desperate to find her a home, so that she could spend what time she had left knowing what it was to be loved and cared for, and, thankfully, Molly's luck finally began to change when she was adopted by her new owner in 2010.

When Molly was first adopted she had nothing – not even a name – and had never worn a collar and lead. The only kindness she had been shown was by staff at the rescue centre. Molly's new owner was determined to bring Molly out of the darkness and misery of her former existence, to make the most of the time that she had left, filling every day with love, affection and companionship.

Molly flourished in her new home, blossoming into one of the kindest and gentlest dogs you could ever hope to meet. A firm favourite at many dog shows, Molly's quiet grace and dignity never failed to win over everyone, and she made so many friends, both human and canine.

Despite her initial poor prognosis, Molly went on to live for another four years until cancer once more reared its ugly head. Molly and her owner fought with every ounce of strength that they could muster but, sadly, Molly lost this, her last battle, a battle that she fought with great courage and spirit.

Although sadly missed by her owner, there is comfort to be had in knowing that Molly did not have to fight alone, and that, from the minute she was adopted, she was surrounded by those who loved her.

This chapter is dedicated to the memory of Molly, and also to new beginnings for two very special girls – Barley and Betsie – who have recently embarked on their own journeys, after being rescued from a puppy farm.

Hang on in there, girls, a new and better life is waiting just around the corner ...

6 Scruffts

Another highlight of our show calendar was the annual 'Scruffts' crossbreed competition at Crufts. Exclusively for crossbreed dogs, this competition was run across the country as a series of qualifying heats, culminating in a Grand Final at Crufts, the world's biggest dog show. Crossbreed dogs could enter a variety of classes, including Handsomest Dog, Golden Oldie, and Child's Best Friend.

In 2013, a new class – Best Crossbreed Rescue Dog – had been introduced, where contestants were judged on several different criteria, including appearance, temperament, and general health. We couldn't wait to attend a regional heat to see how Scout would fare in this class.

> ❝ Getting to Scout, the judge was treated to one of his world-famous leans ... ❞

Our closest regional heat was held in the grounds of Rockingham Castle, Northamptonshire, as part of the International Canine Agility Festival, and we pitched up bright and early on the day with Scout in tow, sporting a brand new blue show collar, which perfectly complemented his dazzling blue eye. Scout's coat had been brushed until it gleamed in the morning sun.

We watched patiently as each class was judged. Scout was in top form as he joined a large number of crossbreed dogs in the show ring for the rescue class.

The judge made his way around the ring, stopping at each dog and owner in turn to chat briefly, and hear their rescue story. Getting to Scout, the judge was treated to one of his world-famous leans: not one to let an opportunity for a stroke pass by, Scout pressed firmly against the judge's legs, his tail swish-swishing gently, side-to-side. 'Gotcha' is probably what he would have said if he could!

It must have been a difficult choice for the judge that day: there were so many wonderful dogs, many of whom had survived terrible cruelty and neglect. In all fairness, every dog deserved to win, but there was only going to be one top dog that day. With a sense of anticipation hanging in the air, we awaited for the judge to make his decision.

In reverse order, the names of the top three dogs were called out, and I could not believe it when Scout was declared the winner! We were called into the centre of the showring for Scout to receive his first place rosette to rapturous applause. Scout stood proudly as we were photographed next to the official Scruffts first place winner's board. We were going to the semi-finals!

Impatiently, I marked off the days on the calendar, as the date of the semi-finals drew closer. Finally, we made the long journey to London, a freshly-bathed and groomed Scout comfortably ensconced on the back seat of the car.

The semi-finals were held at Olympia as part of the annual Discover Dogs event celebrating all dog-related matters, and we had allowed plenty of time to browse the stalls there before checking Scout into the main ring for his class.

When it was our turn to walk out into the ring, my nerves began to kick in. What if Scout didn't want to walk out into the ring? Confident though he was, the ring was huge, and flanked on all sides by rows and rows of seating. There was a constant buzz of activity: would he cope? Would I trip over Scout in front of the audience as we walked around the ring? Seeking to give and receive reassurance, I knelt to give Scout a quick cuddle, which was instantly returned. Scout put his paws on my shoulders and smothered my face with licks. That was all I needed – we were ready to go.

As our names were called we made our way out into the ring. I broke into a gentle trot as we were asked to complete a circuit, before taking our place in the final line-up, Scout matching me stride-for-stride with his high-stepping gait, his tail held high; every inch the showman. All my nerves melted away: Scout was loving this!

Once in the line-up we waited our turn as the judge greeted each dog, and had a brief chat with their owner. Unexpectedly, my nerves returned with a vengeance, and I felt my legs begin to tremble and sag. Kneeling on the floor to keep from falling over, I held Scout against me, trying to draw once more on his strength and confidence. I looked out into the crowds to see if I could spot Paul, to try and give me something to focus on, but the sea of faces were just a blur. I honestly thought I was going to pass out there and then.

A gentle nuzzle brought me back to earth, and I looked at Scout, who had lifted his head to fix his one sapphire blue eye on me. Instinctively, he knew that I was struggling. Starting with my left ear, Scout then began licking my face in earnest, the gentle tickling sensation making me smile as he busily worked away, moving from my ear to my cheek, to finish with a flourish when he reached my chin. Tears pricked my eyes as I ruffled the thick fur around his neck. "Thank you, Scouty" I whispered, as I got back up onto my feet and stood tall and proud (well, as tall as a five-foot tall person can do!).

Our turn came to greet the judge. I was asked about Scout's rescue story, and spoke about his poor start in life, and how he had beaten the odds to survive. I was conscious throughout of Scout's presence as he leaned against my legs – his way of spurring me on.

Tension began to build as the judge reached the end of the line, and made her way to the centre of the ring to give the dogs a last look over. The now-familiar Scruffts boards were placed, numbered first to third place.

As in the qualifying heats, the winners were announced in reverse order and Scout's name was called as third-place winner! I could scarcely believe my ears. Tears of joy streamed down my face as we walked forward to take our place next to the Scruffts board to receive our rosette. Never in my wildest dreams did I expect Scout to be placed. I scanned the crowds again, and this time I spotted Paul, his face glowing with pride.

I still reflect on that amazing day with an enormous sense of pride. To be able to take our place in a line-up of incredible rescue dogs, and share Scout's triumphant journey from unwanted street stray to much-loved family member is a memory that I will treasure forever.

And, as for Scout? He was a real trooper throughout our day at the Discover Dogs event. Negotiating the crowds with his usual unflappable style, he posed for countless photographs, gave out many slobbery kisses, was rewarded with a lot of treats, and made many new friends.

I think he would class that as a result.

7 Sparky

When we adopted Scout, we had not ever considered having more than two dogs, and felt blessed to have Izzy and Misty, who had formed a naturally close bond from day one.

Fate, we believe, led us to Scout, as he was destined to be a part of our lives: so it was that we became a three-dog family.

However, fate can be a capricious thing, and you never know what else it might have in store for you. Inasmuch as fate decided that we would adopt Scout, it also decreed that another challenge would soon come hurtling our way at a hundred miles an hour, in the form of Sparky, the problem Whippet.

As mentioned previously, Whippets are my Achilles heel, and I can't pass one in the street without going weak at the knees.

Many Whippets had passed through East Midlands Dog Rescue where I volunteered, and, thus far, I had resisted the urge to smuggle them out. I should've guessed, though, that there would be one who broke down all my defences, and the dog who took a sledgehammer to my carefully-constructed wall of defence was Sparky.

I first met Sparky one day whilst at work in the veterinary practice, when I was called into reception as Sandy from EMDR had brought in a new rescue dog. Sandy was not scheduled to call in that day, so I was instantly intrigued by her visit.

The first thing that caught my attention was the squirming, wriggling bundle of Whippet that Sandy was holding in her arms, who was trying to get to the small crowd of people gathered around him oohing and aahing.

"I've just brought him in to be weighed" said Sandy, nonchalantly, as she deposited the little Whippet into my arms. "This is Sparky."

This was peculiar, as I was pretty sure that Sandy had a set of scales at the rescue kennels ... I gave Sparky a gentle ear rub as he nuzzled in for a cuddle, his whip-like tail thrashing, side-to-side.

He was simply divine: his velvet-soft coat was a rich cobalt blue shot through with pale fawn, giving off an almost iridescent sheen as the sunlight bounced off it. Sparky's overly-large ears appeared to have a life of their own, and, instead of being neatly tucked away each side of his head like a pair of angel wings, they chose, instead, to form an untidy heap on top of his head. But his most outstanding feature were his eyes: blue-green in colour, they sparkled with vitality, and, even then, a flash of cheekiness that I could discern as he fixed them firmly on me – this stranger who was now holding him captive in a firm embrace.

Maybe I was a bit slow on the uptake; then again, maybe I just didn't care why it was that, for no good reason, Sandy chose to bring Sparky into the vet's that day. Re-homing dogs was her life's work, and Sandy was well practised in the art of matching the right dog to the right person.

I asked Sandy how this lovely little chap, who had now begun to nibble at my earlobe, had ended up in rescue.

> **The expression on his face haunted me: there was such a yearning there, a need to be understood ...**

Sparky's story was not an uncommon one, sadly. Purchased when a puppy, Sparky's owners were totally unprepared for the level of hard work and commitment that comes with any puppy, regardless of breed or size. Sparky had obviously been loved and cared for, but had not received the guidance and training he so badly needed.

Sparky's previous owners had described him to Sandy as being out of control, and they felt they could no longer cope with his behaviour: they confessed they were becoming increasingly fearful of him. This poor little dog was barely six months old, and had already been seen by a canine behaviourist to try to tackle his many 'issues.'

'How can a puppy possibly be that bad?' I thought to myself, as Sparky let go of my ear and started to work on the buttons of my cardigan.

Reluctantly, I handed Sparky to Sandy and watched, deep in thought, as she put him back in the van and drove away ...

A new addition?

I couldn't get Sparky out of my mind over the next few days. The expression on his face haunted me: there was such a yearning there, a need to be understood; a desperate need to be helped. I casually dropped Sparky's name into a conversation with Paul one evening. There was a mild flicker of interest from him, but not the reaction I was hoping for, so I began to up the ante a little.

I tentatively suggested that, as Izzy and Misty had done such a sterling job of looking after Scout, maybe it was time for them to step back a little, and let in some new blood to take over the reins.

I left Paul mulling over the idea.

A new companion for Scout would need to be in possession of a strong character, and robust enough to withstand the knocks and bumps that Scout delivered on a daily basis, characteristics and qualities I felt existed in Sparky. He struck me as a dog who needed a job to do to give him a sense of purpose and focus his mind.

I checked with Sandy over the next few days to see how Sparky was settling in at the rescue centre. She reported that he was doing well, mixing with the other dogs, and even finding himself a girlfriend in the shape of Ella, a young Labrador cross.

Sandy described Sparky as a busy little chap, who possessed a Fagin-like ability to pick pockets without being noticed. He had, apparently, already managed to gather together a collection of purloined items, including various pieces of cutlery, and a twenty pound note he had stolen – not just once, but twice – from Sandy's coat pocket!

I next met Sparky at the rescue kennels about a week after he had been taken in.

Magnet-like, I was drawn straight to his kennel. He rushed straight over, and tried desperately to reach me through the bars, pressing himself sideways against the bars of his kennel in an attempt to be fussed and stroked.

This was as much as I could take. I rang Paul, and asked him to come straight over with Izzy, Misty, and Scout. There was something about Sparky that had triggered an overwhelming desire to try and help him.

Paul arrived a short while later, and, after a chat with Sandy, we introduced Sparky to his potential new family.

First introductions went very well. Izzy and Misty were as courteous as ever, exchanging polite sniffs with Sparky in a calm and relaxed manner. We held back a little with Scout, as we didn't want to overwhelm Sparky, but once he seemed completely at ease with Izzy and Misty, we let Scout move closer to him.

True to form, after a cursory sniff, Scout wrapped his paws around Sparky's neck, and we held our breaths as we waited for Sparky's reaction. Being held in a bear hug is not a usual method of greeting, but Sparky, although slightly surprised, held his ground, and stood patiently whilst Scout continued to embrace him.

So far so good

Once Scout released his grip we took them all for a short walk. I walked with Izzy and Misty and Paul took Scout and Sparky. Considering that Sparky had spent most of his life inside, his on-lead skills were surprisingly good, and he walked steadily at Scout's side. Something about Sparky told me there was so much potential inside this little dog that just needed to be unlocked.

The question was: did we have the key? Maybe there was only one way to find out ...

A tornado strikes

Paul and I walked the four dogs back to the kennels, and had a chat with Sandy. Sandy has always been totally upfront and honest about all of the dogs in her care, and made no secret of the fact that she thought Sparky would be a 'bit of a project.' She, too, saw a lot of good in him, however, and his need for a purpose.

On the strength of this – and our own gut feeling – we agreed to foster Sparky initially, to see if he could settle into life as a walking companion for Scout. Decision made, we loaded all four dogs into the car, and made our way home.

Sparky was quiet in the car, choosing to stay on my lap, his head on my shoulder, watching the traffic with a great deal of interest, whilst chewing on the strap of my seatbelt. I felt a sense of calm wash over me. Sparky was just a puppy, for goodness' sake, he couldn't possibly be as badly behaved as his previous owners had said ... could he?

Once home, we let the dogs into the back garden to give them space to move about. Sparky was keen to explore, and, nose to the ground, he busied himself with a good sniff around, Scout happily pottering behind him. Once or twice Scout accidentally bumped into Sparky, who stopped in his tracks, gave Scout a curious glance, and then carried on about his business.

One by one, they made their way inside.

Izzy and Misty took up their usual positions on the sofa, and Scout settled himself inside his covered bed, ready for a peaceful afternoon snooze. Sparky had other plans for *his* afternoon, it seemed ... As if someone had flicked his 'on' switch, he began dashing about excitedly, rushing from one end of the lounge to the other at top speed.

As he ran back and forth, Sparky started to grab at some of the things on the coffee table in the middle of the room. A newspaper, several magazines, and two coasters were the first casualties. Paul and I looked on in bewilderment as paper missiles began flying our way.

Izzy and Misty looked on, horrified, as a magazine torpedoed toward them. Scout retreated further into his bed, maybe sensing that this was probably the best and safest option for now. Once satisfied that the coffee table was now clutter-free, Sparky proceeded to test out the bounce level of the sofa and chairs by throwing himself from one to the other, displaying just how athletic and agile he was. At one stage, following a particularly energetic bounce, Sparky landed on a very surprised-looking Izzy, who – once she had recovered her composure – slid quietly off the sofa and tiptoed upstairs, with Misty following close behind. Flabbergasted, we continued to watch as Sparky hurtled around for a good ten minutes or more, before flopping down onto the floor, seemingly exhausted.

We took the opportunity whilst Sparky appeared to be resting to take off our coats and shoes. I opened the cupboard door to put them away, and, as I reached down to deposit my shoes and pick up my slippers, my hand briefly connected with a soft, warm snout. Before I knew what was happening Sparky had grabbed a trainer out of the cupboard and run off with his prize clamped firmly in-between his jaws. A game of chase ensued, and I eventually managed to prise the trainer from Sparky.

By the time I had put this away, however, Sparky had discovered a vase of tastefully-arranged twisted willow twigs on a nearby shelf, and was busy re-arranging them all over the floor. Grabbing one of Scout's favourite chew toys, I waved this in Sparky's face to try and distract him. Sparky pounced on the toy and began gnawing away at it in earnest.

Throughout this commotion Scout slept on, seemingly oblivious to the crazy goings-on around him. Paul and I felt shattered simply watching Sparky in action, so poured ourselves a large drink apiece, and sat down with it in stunned silence.

Peace reigned for a good half-hour until we left the lounge for the briefest of moments. Returning to the lounge, I heard a very distinctive slurping noise, and was met by the sight of Sparky greedily gulping from the glass of fresh orange juice I had recently poured for myself. The half-full glass of beer that Paul had left on the coffee table had been drained of its contents, the empty glass rolling lazily across the floor.

I clapped my hands as a distraction, and Sparky lifted his head from the glass, orange juice dripping from his lips. I'm pretty sure he tipped me a cheeky wink before scurrying off to throw himself onto a cushioned dog bed, where he emitted a large belch before falling asleep.

And that was day one ...

Life with Sparky

Over the following days and weeks, we began to learn how to live with Sparky.

Having Scout gave us one big advantage with Sparky, a skilful thief whose new environment could potentially have offered the perfect opportunity for him to fine-tune his skills. We were already accustomed to not leaving anything lying about that might prove hazardous to Scout, which effectively limited Sparky's ability to steal: if it wasn't around, it couldn't be stolen!

We learned to close cupboard doors quickly, and open them only as wide as was necessary to allow us to reach in and retrieve what we needed. If opened any wider, Sparky would be in and out again in the blink of an eye with yet another new treasure to add to his collection. Newspapers, magazines and catalogues were locked away and low-level shelves were cleared of their contents. Valuable ornaments were secured in a sturdy cabinet.

Our philosophy was that if Sparky could reach it then it would be gone, with no-one to blame but ourselves.

We were aware that Sparky had probably received little in the way of mental stimulation from his previous owner, and it was becoming painfully obvious that he had also been denied attention, which was maybe why he was such a bundle of pent-up energy. With no real outlet for that energy, Sparky had taken to stealing to keep himself occupied: a behaviour which may have provided the attention he seemed so desperately to seek, even if that was just people trying to wrestle their belongings from him. Sparky rarely destroyed any of the items that he stole; he seemed simply to relish the act of stealing.

We knew that Sparky needed to redirect his energy to something more productive and meaningful. Izzy, Misty, and Scout were going to be the making of Sparky – we just knew it.

Sparky loved being in the company of our other three dogs. As self-appointed matriarch, Izzy had already assumed the role of tutor, and Sparky appeared to be a willing pupil. He loved to join Izzy for a romp in the garden, and, together, they would kick up their heels and run riot. Izzy also introduced Sparky to the joys of shredding large cardboard boxes. When she was a pup, Izzy also had boundless energy, and was always up to mischief. To help her work off some of her surplus energy, we would place a large, empty cardboard box in the garden (being careful to remove staples, sticky tape, etc, first), peeling back some of the top layers of cardboard, leaving strips of loose material as enticement for Izzy to grab at. She would spend a happy hour or two tugging away at the loose bits of cardboard before eventually shredding them into tiny pieces, then would take great pleasure in dismantling the box, bit by bit. This activity was always guaranteed to wear her out for a good few hours afterward. Fortunately, Sparky also thought that box-shredding was a wonderful way to pass the time, and it seemed an ideal way to help take his mind off stealing.

❝ Scout would go out onto the garden, and bark outside the back door until Sparky rushed out at top speed ... ❞

We began Sparky's basic training straight away, and soon discovered that he was a quick learner, and, more than anything else, desperately keen to get it right. Sparky's ability to walk alongside Scout seemed to come naturally. Sparky was so light and agile that, weaving and skipping around Scout as he constantly shifted direction, became part of Sparky's day, that he adapted to straight away.

And, oh, how he loved his bumbling, clumsy brother! Sparky loved to tease Scout by sneaking up behind him and grabbing Scout's tail, before running off at top speed. A very surprised Scout would whip round to try and catch his assailant, who, by this time, had nimbly skipped away. Scout

would then drop into a play bow, and bark for Sparky to come back, which Sparky would do, dancing around Scout as if daring Scout to catch him. Scout had little chance of doing this, of course, but loved the game, nevertheless.

Scout also loved to chase Sparky around the garden. Scout would go out onto the garden, and bark outside the back door until Sparky rushed out at top speed to join him, often leaping over the top of Scout as he did so. Scout could obviously feel the whoosh! of air as his companion sailed over him, and this seemed to help Scout determine which direction Sparky was headed in.

The pair also loved to play tug-of-war with soft toys. Sparky would present Scout with a toy, holding it in his mouth just in front of Scout's face. Sparky would stand perfectly still until Scout had managed to grab hold of some part of the toy, and the tug-of-war would commence. Scout often lost this little game, but it didn't stop him playing.

Sparky had a patience and tolerance with Scout quite out of character with the way he lived his life at breakneck speed. It was as if Sparky could switch to slow motion when he was with Scout, and their every interaction appeared to be carefully thought-out and planned. We felt that Sparky was the right dog for us from day one.

We had promised Scout companionship, and that is exactly what he got with Sparky. Our family was now complete.

8 Family dynamics

Our lives continued to tick along, sometimes in a very haphazard manner, due, for the most part, to Sparky and his ability to inject a little chaos into every day. His boundless energy and curiosity often got him into scrapes, but he was such a loveable rogue, we couldn't help but forgive his indiscretions. Sparky's bond with Scout was plain, and it was heart-warming to see how much Sparky cared for and looked out for his clumsy older brother.

The dynamics between Izzy, Misty, Scout, and Sparky were becoming firmly established. Izzy continued in her role as team leader, overseeing them all, stepping in with a stern reminder of a strict code of conduct whenever Sparky forgot his manners. Misty was, as ever, Izzy's shadow and soul mate, quiet and unassuming. Izzy had begun to relax a little when it came to supervising Scout, almost as if she had handed the reins of responsibility to Sparky, whose turn it was to keep a watchful eye on Scout. When Sparky's brain switched to caring mode he forgot about stealing things, or performing the wall of death around the lounge, and it also precluded him from participating in one of his favourite pastimes – 'cat tobogganing.' This involved him dashing upstairs to where one of the cats was sleeping peacefully in their cat bed, grabbing one end of the bed containing the sleeping cat, and dragging it down the stairs. Nine times out of ten, Sparky was so fast the poor cat had no chance to wake up and vacate the bed, so was unceremoniously bumped down the stairs.

> **" ... in our efforts to strengthen the bond between Scout and Sparky, we had inadvertently created a monster ... "**

If the cat was still in the bed by the time everyone reached the bottom, Sparky considered this a huge bonus, and would rush off to fetch a toy to deposit in front of the, by now very startled, cat, as if presenting her with an award for endurance! Cats, of course, do not take kindly to any perceived loss of dignity, and Daisy and Jasmine would assume nonchalance in an endeavour to save face as they silently extricated themselves from the car crash that was their bed, before strolling away.

We felt that, overall, we were doing well with our little canine family, developing well-established routines and team harmony. To Paul and I, pretty much everything in the garden seemed rosy.

However, in our efforts to strengthen the bond between Scout and Sparky, we had inadvertently created a monster ...

Seeking help
Sparky had been with us for about six months when we began to notice a subtle change in his behaviour whilst he was out walking with Scout. From day one, Sparky had chosen to walk at Scout's side: it seemed to come

as second nature to him, and so we were happy to let him carry on.

However, we began to realise that maybe Sparky was beginning to take the role of Scout's protector a little too seriously.

When Sparky spotted an approaching dog whilst out walking, he would place himself between Scout and the approaching animal, slowing his pace, and staring intently at the dog until they had passed. For the most part, Scout was completely oblivious to this new behaviour, although, once or twice, he did find himself being suddenly shunted sideways when Sparky took up his new offensive stance.

We didn't feel that this new behaviour was anything to worry about, but problems began to develop if the approaching dog showed an interest in us, be this in the form of a casual lean over for a sniff, an excited bark, or a more enthusiastic pull and strain on the lead to get closer. Behaviour like this seemed to trigger Sparky into more direct action, as if every approaching dog was a potential threat to Scout, and it was his job to drive away this perceived threat.

Sparky's response began to escalate at an alarming rate, to the point where, emitting a banshee-like wail, he would hurl himself at nearly every dog who came near, baring his teeth and growling fiercely. Poor Scout became totally bewildered by all the commotion, and would occasionally fly into an uncharacteristic panic, twirling around and barking furiously, perhaps afraid that something

catastrophic was about to happen, and trying his best to protect himself. Not even Izzy seemed to want to step in to defuse this problem. We soon came to realise that we needed help to try and restore peace and harmony.

That help came from a lovely lady called Marie Miller, who ran a local dog training centre. I had heard nothing but positive reports about Marie, and her insight into canine behaviour and training. After dropping her a quick email I arranged for us to have a one-to-one session with Sparky and Scout to try and get to the bottom of Sparky's reactive behaviour.

When seeking behavioural help, it is vitally important to ensure that the behaviourist you see is in possession of the appropriate credentials. Almost anyone can set up as a behaviourist these days, and the world of canine behaviour therapy is flooded with 'dog whisperers' and self-proclaimed 'experts,' who are more than happy to dole out advice and training techniques with little or no training. Canine behavioural advice is ever-changing and, sadly, · many outdated training techniques are still recommended by poorly-informed individuals, who lack the necessary knowledge and experience to make accurate assessments of different behavioural issues. Some of the techniques can even make behavioural problems worse.

I had done my homework thoroughly, however, and was confident I was making the right choice in Marie. She could help me to understand what was motivating

Sparky's current behaviour, and how best to help him.

Brotherly love

We arranged to meet up at Marie's training studio for our session. I had given her a brief outline of the problems we were having with Sparky, and she had kindly agreed to help.

Initially, Marie wanted to meet Scout and Sparky to observe their normal behaviour, so once she had been treated to the standard enthusiastic greeting by them both, I let them off their leads in the secure studio to let them investigate. Scout immediately made his way over to one side of the studio, and began to mind map his new surroundings, steadily working his way around, nose down, sniffing intently. Sparky, on the other hand, rushed straight over to a pile of toys.

Marie sat watching them for a while, as we chatted about Scout and Sparky and their relationship. Marie commented that although Sparky had chosen not to follow Scout, he did look over at him periodically as if checking that he was okay. I had to admit that, although I was aware that Sparky did like to be close by Scout, I hadn't realised just how great that need had become.

Marie eventually called the two of them back to us and produced a clicker – a small plastic device containing a metal strip that, when depressed, made a short, sharp, clicking noise (clicker training is a positive, reward-based training technique).

Sparky responded instantly to the clicker's noise: ears up, head tipped to one side, he looked at Marie with intense curiosity. Marie rewarded Sparky with a tasty treat for responding to the sound. Scout, on the other hand, did not (and, to date, never has) responded to the noise, although his ever-twitching nose quickly honed in on a treat. Marie repeated the clicker noise several times, and each time that Sparky responded by looking her way he was rewarded with a treat. Marie was satisfied at this point that using the clicker could be a way to help Sparky with his behavioural issue.

As well as trying to modify Sparky's behaviour, Marie was also keen to get to the bottom of why he was reacting in the way he was. She took us outside to an enclosed area to introduce Sparky and Scout to two of her own dogs, and assess how they each reacted.

The first dog Marie brought out was Wispa, a very calm and quiet Labrador-Poodle cross. Marie walked Wispa slowly up and down, asking me to do the same with Sparky and Scout so that we passed by one another once or twice. Sparky regarded Wispa with mild interest, turning to look at her as he walked past. Sparky had placed himself between Wispa and Scout, but his body language indicated that he was feeling confident Wispa did not pose a threat to Scout.

Sparky's mood changed, however, when Marie brought out her second dog, Poppy, to repeat the exercise. Poppy, a working Cocker Spaniel crossed with a Poodle, was a lot bouncier and excitable than Wispa had been, and Sparky's entire body posture changed as soon as he spotted her. Dropping low to the ground, he tensed, and fixed Poppy with an almost unblinking stare. As Marie walked Poppy within a few feet of Sparky and Scout, Sparky started to growl and bark, and tried to lunge at her. Scout began to show signs of confusion, spinning around and barking, as he had previously done.

Marie had now witnessed first-hand just how reactive Sparky could be, and we returned inside to discuss what had just happened.

Marie felt that Sparky had escalated his job as Scout's protector to a higher level. She explained that Sparky had taken on this role at a very young age, and was genuinely doing the best he could to try and keep Scout safe. Due to Sparky missing out on so much socialisation as a puppy, he was essentially creating his own rules about how to handle a potential threat, never having received any guidance early in his life about how to interact politely with other dogs whilst out on a walk. Marie empathised with Sparky, who was very confused, and maybe a little bit anxious about not being able to fulfil his role.

Added to this was Izzy's lack of intervention when Sparky began to react in this way. We had relied on Izzy to help keep Sparky on the straight and narrow, but she seemed rather flummoxed by the recent changes in his behaviour. This could also have led to increasing anxiety on Sparky's part, Marie thought, as he had no safety net to protect him. Sparky, it seemed, was feeling a bit out on a limb ...

Marie explained that Sparky had learnt that behaving in an aggressive way toward certain dogs meant we would simply move away to a safe distance from the other animal, which was exactly what Sparky wanted. We were inadvertently rewarding and re-enforcing his reactive behaviour by doing this, and, in his mind, Sparky thought that he was doing a grand job of protecting Scout from potential harm.

Being constantly on high alert during every walk was not good for Sparky in the long-term, though, and our fear was that this aggression would become more commonplace, and even increase in intensity if it remained unchecked. Marie felt that Sparky had taken on a lot of responsibility at a very young age, and was struggling to cope when faced with certain challenges when out on a walk.

Realisation quickly dawned that what Marie was saying was exactly right. We had given Sparky a project from day one – Scout – and thought that, by channelling Sparky's excess energy into looking after Scout, this would help with some of his other issues. The reality, however, was that although Sparky was desperate to fulfil his given role, because of his lack of socialisation, he was beginning to feel out of his depth whenever he perceived that Scout was under threat from an unknown dog. Sparky was lacking the social skills required to be able to recognise the difference between a potential threat and a dog who just wanted to say hello.

Knowing this helped explain why Sparky reacted to certain dogs and not others. He rarely reacted to calm, steady dogs who walked by without even a glance his way, as he did not consider these dogs to be a threat, but this was not true of bouncy, lively dogs who came rushing over, not allowing him time to accurately assess the situation. Sparky would also react to short-nosed dog breeds – Pugs, Boxers, Bulldogs, etc, as these tended to breathe quite noisily, which Sparky may have interpreted as growling or snarling. Short-nosed breeds often naturally expose their lower teeth which, again, Sparky may have interpreted as an aggressive facial expression.

I felt so sorry for Sparky: after all, he was just trying to do his best. We had never admonished him for behaving the way that he did, because we knew there had to be a reason for it, and now that we had clearer insight into what was motivating Sparky to exhibit aggressive behaviour, we could begin to help him. I knew it wouldn't be a quick fix, and that it would take time and patience, but I felt sure that we could work with Sparky to help him overcome his problem.

Click and reward
And so, the hard work began in earnest. Following advice from Marie we equipped ourselves with a clicker, and never went anywhere without a pocketful of treats.

Starting in the house with no distractions we used the clicker at random times when Sparky was with us – every time we clicked, he received a treat – and we repeated this until the action was firmly imprinted in his brain: the sound of the clicker meant a food reward was coming.

Once Sparky was comfortable with the idea we moved out onto the garden, walking him up and down on a loose lead. If Sparky walked quietly at our side we clicked and rewarded, marking the calm behaviour as that

which earned him a reward. We moved on to taking the clicker on walks with us, trying to pick times and places where there were no other dogs to challenge Sparky. We practiced loose lead walking, clicking and rewarding calm, steady progress.

Sparky continued to be very focused on the clicker, eagerly anticipating the treat that followed each click: as a result, the constant scanning for potential threats reduced. As well as using the clicker to reward calm walking we also constantly praised him for doing well. Throughout this training, we continued to walk all four dogs together. Scout was still oblivious to the clicker and what it represented, but was happy enough to snaffle the odd treat or two, along with Izzy and Misty.

With our confidence in Sparky growing we began to walk in areas where we were likely to encounter other dogs, though, when we did, were careful to remain at a reasonable distance to ensure that Sparky stayed calm and relaxed.

One of the key aspects of positive reward-based training is to never set up a dog to fail, because the more times that he or she succeeds the more likely it is that the new behaviour will stick. Admittedly, we had our fair share of failure at the start as Sparky was adept at catching us out. Sometimes he would give a passing dog a cursory look, and continue on his way, only to suddenly emit an ear-piercing shriek and lunge – front legs flailing wildly as he strained against his lead – toward the other dog. Of course, raising your voice at a dog who is barking and shrieking can seem to the dog as if you are joining in, and serve to exacerbate the situation by inadvertently re-enforcing the unwanted behaviour. We learned to simply wait until Sparky stopped throwing himself about, when we could continue on our way, clicking and rewarding once calm and order had been restored.

I must confess that sometimes I felt like a Maypole when Sparky wrapped his lead around my legs several times as he whirled and twisted about, especially as Scout tended to do the same. Misty took to retreating behind my legs at any sign of trouble, and Izzy often ended up in the middle of it all; a somewhat puzzled expression on her face. How I didn't fall flat on my face during these frenzied episodes, I will never know. We weathered the storm each time and waited patiently for calmer waters.

With dogged determination, we stuck to our plan, trying to keep Sparky within his comfort zone whenever we encountered another dog, rewarding relaxed loose lead walking; marking appropriate behaviour with a click and a treat. Some days went splendidly well; on others it was a case of coming home and lying down in a darkened room! But as each week went by Sparky's reactivity began to diminish.

As already noted, there's no quick fix for reactive dogs. Finding the trainers and behaviourists to give the correct advice, and following that advice to the letter, employing only positive, reward-based methods, and taking the time to understand the motivation behind reactive behaviour, is the *only* way to help a dog to deal with the many different emotions that can trigger a reactive response. Sadly, many reactive dogs are labelled aggressive simply because no one has taken the time to fully understand their problems.

> **66 We learned to simply wait until Sparky had stopped throwing himself about, when we could continue on our way ... 99**

It would be foolish of me to say that Sparky is cured. A cure was not what we were aiming for, in any case. What we needed was guidance on how best to manage Sparky's problems, along with clearer insight into what was motivating him to behave in the way that he did ... and that is exactly what we got. Sparky can and will still react sometimes to other dogs when out on a walk. For example, he has a pathological hatred of a dog who lives nearby, who likes to push his nose under his garden fence and bark as we walk by. Sparky seems to find this behaviour extremely rude, and doesn't hold back in making his opinion known. He'll also still react if another dog directs aggressive behaviour toward himself and/or Scout, positioning himself in front of or at Scout's side, behind an imaginary line in the sand – and woe betide any dog who dares cross it!

Ultimately, what we were hoping for when we adopted a companion for Scout was a dog who would look out for him, and would always try to keep him safe. Sparky may always be a bit rough around the edges, but he has the biggest and kindest heart. Even if, sometimes, his actions are a little misguided, his intentions are always good.

g Izzy

Just how we went about fulfilling our promise to guide and support Scout is reflected in the relationship between him and Izzy. Izzy was Scout's guiding light ... and he adored her. Despite her headstrong, madcap ways, Izzy was a natural leader, who established herself very early on as Scout's personal trainer. Whatever she did; wherever she went, Scout would follow suit. Izzy was already accustomed to having an adoring disciple glued to her side in the shape of Misty, who was perfectly happy to be Izzy's little shadow, always two steps behind her older (and often not so wiser) sister. Adding another protégé to Izzy's fan club seemed like a natural progression.

From day one, Izzy kept a watchful eye on Scout, sometimes in a very obvious way, although she could also be very subtle. If Scout was ever alone in the garden for more than a few minutes, Izzy would dutifully trot outside to check he was okay. If Scout became a little over-excited, she would give him a polite nudge with her nose to remind him of his manners. On days Izzy felt a twinge of devilment, she would stand at the top of the stairs and bark for Scout to come up and join her in a spot of duvet shredding.

With the addition of Sparky to our family, it seemed as though Izzy had her paws full, but she thrived on company and, once she had gotten over the initial shock of the effervescent whirlwind that was Sparky, Izzy once again stepped up to the plate and became a pivotal figure in the taming of our rambunctious little hound.

The one area in which Izzy refused to have any input was dealing with Sparky's reactivity and over-protectiveness of Scout, but maybe this was because her intuition was telling her that this particular issue was beyond even her capabilities, and was probably best left to the experts, with which I am inclined to agree.

We expected a long and healthy life for Izzy, an incredibly fit dog who rarely had a day's illness. We could not possibly have known that we were heading toward the darkest and saddest of times with our beautiful girl ...

Fighting demons
It began one morning toward the end of summer 2014.

I had let the dogs into the garden as usual to go about their business, and stood watching them for a while as they pottered about. Izzy was usually the first out, closely followed by the rest of the gang, but this morning she seemed a little slower than usual, and appeared to stumble a little as she stepped down from the back door. I continued to watch as she slowly ambled across the garden, and noticed that her tummy looked a little more rounded than usual. Izzy had a very athletic build and carried not an ounce of extra fat anywhere, so, troubled by this subtle change, I booked an appointment for her to see Viv, her vet.

Viv gave Izzy a thorough examination from nose to tail, and

agreed that her tummy was a little distended. Based on his initial examination, Viv recommended an ultrasound and blood tests to investigate further. Izzy was a model patient as the ultrasound probe was passed gently over her abdomen to check her internal organs, standing calmly and patiently throughout. I kept my hand on her back, stroking her, and whispering quiet words of reassurance into her ear.

Once the ultrasound scan was complete, Viv removed the probe and told me what he had discovered. The news literally took my breath away. Izzy's liver was in trouble; fluid was accumulating in her abdomen because her liver was failing to function properly. Subsequent blood tests revealed that Izzy's liver readings were abnormally high: she was heading into liver failure.

I cuddled Izzy against me as Viv talked through the list of potential causes of her liver failure, and I could not hold back the tears as I listened to what he had to say.

At best, he advised that Izzy could have contracted a severe liver infection that could potentially respond to treatment, although his overwhelming concern was that Izzy's liver was already showing signs of significant damage. The worst-case scenario was that Izzy had liver cancer, which, if this was the case, meant the prognosis was grave.

I couldn't take it all in – the prospect of losing Izzy was almost too much to bear. She was only nine years old, and had always been such a healthy dog.

Viv recommended that we start Izzy on a number of different medications to help her liver try and fight whatever it was that was causing it to fail. Armed with a vast array of tablets and medicines, I took Izzy home, my head spinning as I tried to absorb and process all the information I had been given.

Over the following weeks, we fought the battle of Izzy's 'liver demons,' as we decided to call them, in a desperate attempt to halt the progression of her illness. Despite everything we tried, Izzy deteriorated at a rapid rate.

Gone was our bossy, exuberant team leader with her agile mind and adventurous spirit. Izzy began to tire easily; her once voracious appetite dwindled to nothing. Her bouncing gait became a laboured shuffle, and she also began to stumble from time to time as she walked. Eventually, Izzy was unable to jump onto her favourite resting places, and needed to be lifted there. She also began to experience periods of confusion, during which she would stare off into space, and sometimes she would even fail to recognise those who she loved so dearly, and who loved her in return. It broke our hearts to watch our girl fade away before our eyes. Every day we looked for even the smallest sign of improvement, but in vain.

Blood samples and scans over the next few weeks revealed that Izzy's liver was continuing to fail, no matter what we tried. We knew that the time was fast approaching when we would have to make the decision to let our Izzy go.

Saying goodbye
One bright and sunny day in October we knew that it was time to say goodbye to Izzy.

We spent that last morning together quietly. I lay with Izzy sleeping peacefully at my side, reliving in my mind the many happy memories of our time together. Those giddy early days when she was a puppy; her liking, then, for ripping up carpets to make comfy little nests out of them. Coming home at night to see her capering around in a snowstorm of carpet pieces and shredded underlay was a regular occurrence. Izzy had such a fondness for meeting people, too, wherever we were, and would rarely let anyone pass by without reaching out to nudge them with her nose.

As she grew Izzy matured into a stunning beauty, with a stature and grace that belied the cheeky, mischievous character that lay beneath. Izzy was only just approaching the autumn of her life and, just as autumn fades into winter, so, too, was our Izzy. If love alone could have cured her, Izzy would have been restored to us in full health, but, with heavy hearts, we knew we had reached the end of our journey with Izzy, and that it was time to let her go.

" She was bathed in the warmth of the sunlight that she loved so much. My heart shattered into a million pieces that day. Our guiding light had gone forever ... "

Viv the vet called at our house that day. As he knocked on the door Misty, Scout, and Sparky rushed out to greet him, Izzy climbed slowly off her bed and followed them, stumbling once or twice as she walked. When she saw Viv, Izzy tipped her head quizzically to one side, as if trying to remember where she had met him before. There was a brief flash of recognition in her eyes before she turned and ambled back to her bed. With a loud sigh, Izzy closed her eyes

and dozed off. Viv came and sat at her side, and spoke to her gently and with calm reassurance. Izzy was aware of his presence but she didn't stir from her resting place.

I held Izzy in my arms as she left this world peacefully and with dignity. I felt her life slip away as gently as a soft breeze that blows through the trees on a summer's day. She was bathed in the warmth of the sunlight that she loved so much.

My heart shattered into a million pieces that day. Our guiding light was gone forever, but I take great comfort in the knowledge that Izzy's legacy lives on in Misty, Scout, and Sparky. Without Izzy, we may never have had the confidence to adopt Scout. Without Izzy, we would never have had the courage to take on a problem dog like Sparky.

I remain ever-thankful we were able to share our lives with Izzy, who taught us so much and gave us so much, too

Learning to cope

Our lives changed in many ways after losing Izzy, and we all struggled in one way or another during the months that followed. Misty, Scout, and Sparky were very much affected by the loss of their leader, and it became clear that Izzy was the glue that held them all together: without her there to call the shots, all three seemed to lose their sense of order and direction.

Misty, for example, became very withdrawn. Izzy was her soul mate, and although as different as night and day, the two complemented each other beautifully. Sparky was also beginning to really benefit from Izzy's wise ways; she was instrumental in helping to curb some of his less desirable traits.

Interestingly, it had been Sparky who chose to spend a lot of time with Izzy during her illness. For a dog who could usually barely keep still for more than a few seconds, Sparky spent many hours sleeping quietly at Izzy's side during her last few weeks. After she died, Scout became confused; he took to visiting Izzy's favourite spots in the house or out in the garden, where he would stand for the longest time, as if waiting for her to appear beside him. He'd stay outside so long sometimes that we would have to pick him up to bring him back inside. It took some time before Scout seemed to accept that Izzy was never coming back.

Somehow, we managed to muddle through. Losing the strength and guidance of an established and respected team leader hit us all hard, but, one by one, we managed to get ourselves back on track. That would, after all, be what Izzy would have wanted.

I believe that Izzy watches over us. I also believe that she sent us a message to tell us just that, and she did it in true Izzy style.

The unassuming vehicle who carried this message was Sparky. Within forty-eight hours of Izzy's passing, Sparky's eye colour changed overnight from sea-green to a rich chocolate brown: the exact same shade as Izzy's.

Little snippets of Izzy's behaviour manifest themselves now and again in Sparky, which we fondly refer to as Izzy-isms. Sparky's first reaction to running free on a sandy beach mirrored Izzy's exactly; Sparky's loathing of being cold and dirty, and his love of shredding cardboard boxes, all evoke the fondest memories of Izzy. To date, Sparky hasn't ripped up any carpets, but that's probably because we were one step ahead of him on that score, and had our carpets replaced with wooden flooring.

Run fast and free at Rainbow Bridge, Izzy: know that you were always loved and will be forever missed ...

10 Awards and achievements

Scout – the consummate showman that he is – has won countless accolades and awards since the early days when he was a skinny, mange-ridden, unwanted puppy.

We have a tall, glass-fronted cabinet in our lounge which displays many special awards, trophies, and an assortment of memorabilia that Scout has managed to accumulate over the years. Each one of the awards in the cabinet represents a special memory for us, and has its own story to tell.

Four of these awards, in particular, I hold very dear to my heart as I feel that they represent Scout as the remarkable little dog that he is, as well as the impact he has had, not only on our lives, but on those of so many others. They also remind me, every time I look at them, of the promise I made to Scout that his name would be widely known, and he would always have a purpose in life.

The first of these awards is a framed certificate, presented to Scout at an official party thrown for him and his fellow comrades by the RSPCA at its UK headquarters in Horsham, West Sussex, in 2015.

As one of the largest animal welfare charities in the UK, the RSPCA launched Ruffs in 2014: an alternative, online dog show intended to celebrate all that is good about dogs, no matter what breed or background.

Choosing to focus on the health and happiness of the dogs who were entered in the show, rather than on how they looked or conformed to a certain standard, Ruffs seemed like the ideal platform for Scout to show the world who he is and where he came from.

> **❝ Scout had, it seemed, touched the hearts of hundreds and hundreds of people ... ❞**

The Ruffs competition was divided into several different categories, including Happiest Hound, Golden Oldie, and Best Transformation (a category for dogs who had managed to overcome terrible experiences of neglect and cruelty). But the class that really caught my eye was that celebrating the 'perfectly imperfect': those dogs who would never conform to any kind of official standard, but who, in the eyes of their owner, were perfect in every way. The class seemed almost tailor-made for Scout, and, as I submitted my entry, it was with fingers and toes firmly crossed.

I was over the moon to be contacted by the RSPCA a short time later, and told that Scout had been selected as one of the five finalists in the Perfectly Imperfect class. The overall winner of this class was to be decided by public vote, and as voting began, I was overwhelmed by the response. Votes for Scout flooded in, as well as countless messages of support and goodwill. Scout had, it seemed, touched the hearts of hundreds and hundreds of people with his story of triumph over adversity, proving once again that disability is no barrier when it comes to achieving goals.

Paul and I stayed up until midnight on the evening that voting closed, and celebrated in style with a mug of hot chocolate each when Scout was announced as winner of the Ruffs Perfectly Imperfect 2015 class. Scout joined in the celebrations with a selection of his favourite biscuits.

Scout thoroughly enjoyed himself at the awards presentation at the RSPCA's headquarters, and even managed a celebratory leg-cock against the side of the buffet table during the proceedings.

Perfectly Imperfect – two words that really do say it all!

Special memories

The second item in our cabinet is a large and very grand-looking trophy. Constructed from marble and gold metal, this trophy was awarded to Scout by Bill and Marie, a lovely couple, who had followed Scout's story almost from the start. We were presented with this award one summer whilst at a dog show, and it came as a complete surprise to us to be invited into the show ring to receive it.

Bill and Marie had the trophy commissioned for Scout, and it bore a small engraved plaque which read 'Best Rescue Dog 2014/15 Scout.' A simple gesture of kindness and thoughtfulness that will never be forgotten.

A delicate silk Peony sitting in a glass vase occupies a small corner in the cabinet. This award holds some very bittersweet memories of a show that we attended, along with our friends during the autumn of 2014.

We had arranged with our friends to spend the day at the show, knowing it was likely to be the last time that two of the dogs in our group would be with us. The two dogs were our very own Izzy, and our good friend, Fay's, rescue Greyhound, Hero.

Hero was adopted by Fay and Darrin prior to them owning their current rescue dogs Bliss, Shadow, and Mila. Hero had been taken into kennels once his racing career had ended following a nasty injury to his foot. Whilst in kennels he had damaged the end of his tail, which was bandaged in an attempt to protect it whilst it healed. Sadly for Hero, the bandage was left in place far too long, and caused irreversible damage to his tail, necessitating a full tail amputation to halt the progression of a nasty infection that had set in.

Fay met Hero whilst he was receiving treatment for his injuries, after he was signed over to East Midlands Dog Rescue, and was instantly smitten. Here was a dog who had suffered a great deal of discomfort and pain, yet the trust in his eyes spoke volumes about his ability to forgive, and the understanding he had that there were people who would do whatever it took to help him recover.

Izzy had been fighting liver disease, which we knew, in our heart of hearts, was a fight that she could not win. At the same time that Izzy fell ill, Hero had been diagnosed with Osteosarcoma, a devastatingly cruel bone cancer that claims the lives of so many Greyhounds. Just like Izzy, Hero fought his illness every step of the way, but Fay knew that his time was growing short, and that this day was all about sharing the precious little left to him.

We pasted on our best smiles and fought back the tears: this was going to be a day of celebration, not sadness.

61

It was a beautiful day for a show. Warm sunshine and a cloudless sky helped to lift our spirits, just a little.

Izzy and Hero were settled comfortably inside a tent we had set up for them. Izzy lay buried under a tangled pile of furry bodies, comprised of Misty, Scout, and Sparky, whilst Hero snoozed quietly nearby on his specially-padded bed.

We hadn't intended to enter many show classes that day – we were more than content to sit and watch the activity going on around us – but there was one special class that had caught our attention.

In memory of the judge's own much-loved dog, Sasha, there was a class for dogs with special needs, be these due to illness or disability. I had entered Scout in this class, and Fay had decided that it would also be Hero's last appearance in a show ring. Side-by-side, slowly we walked into the ring and took our places. Hero's bed was brought out for him, and he was quite content to lie down and relax on this.

The judge made his way round to ask each of us about our dogs and their background. I could see Fay trying so hard to control her emotions as she explained to the judge about Hero's condition; how, despite exploring every avenue, and the best possible treatment for Hero, the cancer was continuing its relentless march, and now there was nothing more that could be done.

I can only imagine how difficult the judge's decision was that day as every single dog in the class was battling to overcome their own difficulties and challenges.

Eventually, he made his choice and announced the class winners.

Third place went to Tyco, a handsome black Lurcher who had lost a front leg, but could still run at lightning speed, and second place went to Scout for his ability to always remain positive, despite everything life had thrown at him. Overall class winner, and a fitting tribute to an incredibly special dog known affectionately to his friends as Big H, was Hero for fighting his battle with so much courage and dignity, a gentle giant with the kindest of souls.

In the days that followed Fay and I comforted and supported one another as we began to come to terms with what lay ahead for us both.

Izzy and Hero lost their fight for life within twenty-four hours of each other, and I like to think that Izzy lit a path for Hero to follow on his final journey, to be reunited with all the friends who passed before them.

I look at the Peony today, and because of the memories that it stirs, can still smile through the tears.

Lastly, in our glass cabinet there is a small, silver trophy. The trophy stands only a few inches tall, and is, I admit, nothing much to look at: its silver badly tarnished and with chips and scratches on the base.

But the word engraved on the plinth requires no explanation – Hero – and Scout is mine, as well as my inspiration and my rock. He keeps me focussed, and gives me the confidence and ability to believe in myself.

I guess if Scout was to make his own promise to me it would be that of life-long companionship; being there when I need him, and making me smile, no matter what.

11 Holidays

Holidays: a time to relax and unwind together as a family, explore new places together and create great memories.

When looking for places to holiday we had a long list of requirements, most of which revolved around the needs of our dogs. Dog-friendly beaches, walks, and eateries were always at the top of our list.

After adopting Scout, another consideration was the *type* of accommodation that we stayed in, as we had to ensure that Scout would be able to safely move around wherever we took him. Gardens had to be relatively clutter-free with easy access, and garden furniture, planted containers, and overstocked borders were all potential hazards. We had to consider the internal layout of the property, too – no open staircases for us; no cosy open fires or log burners: all a potential risk to Scout's safety. Properties with a simple layout that Scout could easily mind-map were what we sought.

> **❝ It was Scout's first holiday away from home, and we were a little worried ... ❞**

One particular summer we were offered the chance to stay in a log cabin owned by friends, situated in Trawsfynydd, a small village nestled deep in the heart of Snowdonia National Park in Wales. As far as our exacting list of requirements went, the cabin and its location seemed to tick every box. We had never been to Wales, and did not know what to expect, but what we found far exceeded our expectations.

And so began a love affair with Wales that is as strong today as it was the first time we visited there.

Our first stay at the cabin with Izzy, Misty, and Scout occurred in 2012, and, in memory of Izzy, I would like to share this with you.

It was Scout's first holiday away from home, and we were a little worried about how he might cope away from his usual familiar surroundings.

On arrival, Izzy and Misty set out to explore their temporary home, and seemed more than happy with their new accommodation. When it was Scout's turn to give the place the once over, we took a little more time. Keeping him on his lead, we walked with him from our car to the small set of wooden steps that led up to the front door of the cabin. Following the previously taught 'step' command, Scout carefully climbed each step and entered the cabin, his whiskers and nose twitching in their usual inquisitive way.

Once safely inside the cabin, we unclipped Scout's lead to give him the freedom to move around inside, and create his mind-map. Which is precisely what he did, slowly and steadily nosing his way around his new surroundings.

The cabin itself was quite small. An enclosed veranda led into a compact kitchen area with an adjoining lounge, whilst two bedrooms and a shower room completed the layout. Room by room, Scout checked out all of the nooks and crannies before hopping onto one of the beds and settling down for a snooze. We'd obviously made the right choice for Scout, and he had given the cabin his very own seal of approval.

Happy times!

Over the next few days, following recommendations from the cabin's owners, we embarked on a mini tour of North Wales in all its beauty and majesty, starting with a day trip to Barmouth, described as one of Wales' most popular seaside resorts.

Izzy, Misty, and Scout had the best of times zooming about on the dog-friendly beach, although Scout was not *too* sure about the feel of the sand beneath his paws, initially, and couldn't work out why his feet kept sinking into the soft, powdery stuff beneath him. He'd lift his front paws, one at a time, and vigorously flick them, showering himself with sand, which then made him sneeze.

Scout was also a little confused by the sea at first. Rushing to the water's edge, perhaps honing in on the sound that the waves made as they crashed on the shore, he ran straight in, splashed about, performed a few spins ... then proceeded to drink the sea water! Getting a good taste of the salty water, he quickly spat it out and retreated to the beach.

As well as the glorious beach at Barmouth, we walked on the beaches at Harlech and at Black Rock Sands, which were just perfect for dog walkers like us. The beaches were usually quiet, and stretched as far as the eye could see. Here, we could relax, as we knew that Scout could safely run off-lead without fear of colliding with or falling into anything.

We also discovered some amazing walks in Wales. Spending a day in the charming picture-postcard village Betws-y-Coed, we completed an arduous walk to Llyn Elsi, a mountain lake hidden amongst the fir-lined woodlands of the Gywdyr Forest. The views looking down onto the lake were simply breathtaking.

Sitting at the foot of Snowdon is the village of Llanberis, where we spent an interesting morning looking around the old slate museum, soaking up some Welsh history, before picking up the trail for another lake walk, this one a circular route around the crystal-clear waters of Lake Padarn.

This walk was particularly challenging for Scout, as some of the paths were incredibly steep and narrow, sharply falling away at one side into deep, fern-filled ravines. A different mix of terrain made up the walk – from narrow woodland trails and old railway beds, to winding country roads flanked on each side by gorse and heather.

For the duration of the walk we kept Scout on his lead and harness, which enabled us to carry him along some of the trickier paths. Scout didn't seem to mind becoming airborne now and then.

Scout's biggest test on this walk was climbing the many wooden stairs that led up to the old quarry hospital, but then all of us needed to take a bit of a breather after tackling them.

The hospital was built in 1860, and closed in 1962, to be reopened many years later as a visitor centre. Serving the quarry workers of the Dinorwig slate quarry for many years, the hospital is now a museum, housing some of the original (and occasionally quite gruesome) equipment from those bygone days. The museum allowed dogs inside it, which meant we could all go in and have a good look round. The view from in front of the hospital was magnificent, looking out across the lake with Snowdon as the perfect backdrop. On a clear day, you could see for miles.

To experience the full beauty of Snowdonia National Park during our stay, we took a train ride on the Ffestiniog and Welsh Highland Railway – another first for Scout, who was a little unsure as the train rumbled and rattled its way out of the station at Porthmadog. Lots of reassurance and gentle ear rubs helped calm him, and he eventually fell asleep on my lap. Misty was rather discomfited after falling off the seat when the train braked suddenly, whilst Izzy, on the other hand, loved the experience, and spent the entire journey looking out of the window at the spectacular scenery.

One of our favourite places to visit was Beddgelert, a quaint little village that nestled amongst the mountains of Snowdonia, and steeped in mystery and legend. It's named after Gelert, the faithful hound who belonged to Llewellyn, prince of North Wales during the thirteenth century. Gelert's grave can be found near the banks of the Glaslyn River, and the tombstone erected there tells the story of this brave hound, tragically slain by Prince Llewellyn because he believed that Gelert had killed his infant son. In fact, Gelert had saved his son's life by killing a marauding wolf, and the child was subsequently found unharmed. It is said that the prince was so full of remorse for what he had done that he never smiled again.

You can still visit Gelert's grave today, near to which, in the ruins of an old stone cottage, can be found a bronze statue of this noblest of hounds.

We tended to spend the whole day in Beddgelert whenever we visited, and, in the morning, would take a gentle stroll along the path that ran alongside the Glaslyn River to pay our respects at Gelert's grave. Izzy and Misty loved to stop along the way to paddle in the shallow waters of the river, whilst Scout preferred to wade in for a good splash about, soaking everyone in the process!

After lunch we would spend the afternoon on a long walk, which took us out of Beddgelert and into the countryside beyond.

Spending most of our days walking gave us all very hearty appetites, and we enjoyed many delicious meals in some of the dog-friendly pubs and cafés situated along many of our routes.

We always received a warm welcome wherever we stopped, and offered towels to dry the dogs with if we had been caught in the rain. Fires were lit to keep them warm, plus many a tasty treat offered and graciously accepted.

After a day spent trekking high into the mountains, or walking for miles along empty beaches, we would return to our cabin to relax before bed, three very contented dogs sleeping peacefully at our feet.

From the first holiday we enjoyed there, a little piece of our hearts has belonged to Wales.

Exploring

Our first trip to Wales without Izzy was during the summer of 2016, a bittersweet holiday for us visiting all of the places we used to walk with her. So many happy memories of sunny days, beach and hill walks – and even trips high up into the mountains on the Ffestiniog and Welsh Highland Railway network – came flooding back.

Izzy loved holidays in Wales. She adored the vast, empty beaches where she could run and run at full speed, and loved to play a game of chase with Misty amongst the sand dunes. Paul and I would watch them darting about, their little faces popping up now and then from behind a dune as each tried to catch a glimpse of the other. Izzy also loved the sea, and would stalk and then pounce on the small waves as they rolled onto the beach, mouthing the flotsam and jetsam as it drifted.

---

66 **... Scout skilfully negotiated several large rocks ...** 99

More than anything else I think Izzy enjoyed our long walks into the Welsh hills and valleys.

As this was Sparky's first trip to Wales, we were keen to revisit our old haunts to see what he thought of Izzy's favourite places.

Beddgelert was somewhere that Izzy always enjoyed, and we returned there to honour her memory. As soon as we arrived we fell right back into our routine of picking up the path alongside the Glaslyn River, as we made our pilgrimage to Gelert's grave.

It soon became apparent that Sparky was no water baby, repeatedly casting disdainful looks at the swirling water that flowed alongside, and opting to walk on the side of the path furthest away from the bank.

Sparky's mood was not helped by the light drizzle falling from the grey clouds above us, but he soon cheered up when we retreated to a nearby pub, where we received a warm welcome from the front of house staff, who quickly ushered us toward a roaring open fire, plying the dogs with tasty biscuits as they did so.

Fortified by the warmth of the fire and a hearty meal, we decided to tackle one of the more popular walks, which started from Beddgelert, and headed out toward Llyn Dinas: the mountain lake that usually offered truly spectacular panoramic views of the surrounding mountains and valleys.

The rain that had plagued us since early morning became more persistent as we left the village and picked up a narrow slate path that ran alongside the Glaslyn River. All of us were fully kitted out with waterproof clothing, having learnt, to our cost, that Welsh weather can be very changeable.

Scout never seemed to be overly bothered by inclement weather, however, and steadily trotted along, happy to be guided by us whenever the path became very narrow, or appeared to sharply drop away on one or other side. If the paths appeared too narrow or dangerous, Paul simply carried Scout over them, and was usually rewarded with a grateful face lick.

Eventually, the slate path we followed began to widen as we approached the Sygun copper mine, an old Victorian working that is now a popular tourist attraction.

Beneath us were the old mining tunnels, from where it was just about possible to hear the recorded tour voices that were played through speakers at different points along the tunnels' length. Scout could hear these voices but was quite unable to work out why they seemed to be beneath him. Now and again, he would stop to stand perfectly still, head tilted to one side in a quizzical way, ears pricked, trying to pinpoint where the voices were coming from!

Eventually, we reached Llyn Dinas, where we sat for a while to catch our breath. The usual stunning views across the lake were obscured by mist and rain, but this didn't dampen our spirits one bit. Well, apart from Sparky's, maybe: he was still sulking about having to walk in the rain.

Whilst Sparky chose to sit with us, Misty and Scout did a spot of exploring, and we watched as Scout skilfully negotiated several large rocks situated at the water's edge. Taking his time, keeping his head low and with his nose almost touching the ground, Scout carefully extended each front paw, testing the ground in front as he went. Climbing on top of the larger rocks, and skirting the smaller ones, he seemed in a happy little world of his own. Every now and then Misty would wander over and give him a gentle nudge, as if to say, 'We're all still here, keeping an eye on you.'

On the journey back from the lake we passed the old copper mine building (now a museum and gift shop), and Paul pointed to a narrow footpath at one side of this. From where I stood I couldn't see where the footpath led, although it did look quite steep as it wound its way uphill before disappearing into the rapidly-descending mist. As we were already pretty much soaked through, I shrugged my shoulders in a 'Why not? Let's follow the path and see where it leads to' gesture. I did briefly stop to read a small wooden sign next to the footpath, which advised against attempting the walk in bad weather. We were experienced walkers, I reassured myself, and it was only a spot of rain and mist. I followed Paul and the dogs up the path.

I lost count of the number of times that I fell over on the walk, slipping and sliding about on the mud, slate, and rocks, landing rather unceremoniously on my backside and slithering downhill at an alarming rate of knots each time. Naturally, I was shown up by Scout, the mountain goat, who confidently hopped from rock to rock with effortless ease. Each time I fell over, Scout would seek me out, patiently waiting next to me until I managed to regain my feet (he also very obligingly washed the mud from my face several times). Paul

was doing admirably well, keeping a watchful eye on Misty and Sparky, as well as managing to hold on to an umbrella, as he made his way up the path.

After we had been walking for a good half hour, the mist became so dense that I could see only a few feet in front of me. I guess this handicap put the rest of us on the same level as Scout, and it was ironic that it was Scout who guided me all the way to the end of the path as I followed the tip of his swishing tail.

Once we got to the end of the path I brushed my rain-sodden, mud-caked hair from my eyes and looked around. I could just about make out the shadowy figures of Paul and the dogs a few feet away, but that was about it, everything else was shrouded in mist. So much for the breathtaking mountain views, I thought, as I peered into the gloom.

Suddenly, I felt the gentle bump of a nose against the side of my leg; it was Scout, checking on me, and yes, his tail was still wagging. View or no view, Scout couldn't have cared less: he was with his family and was having a great time as usual. How could I not smile at his enthusiasm?

To this day, I am not sure whether the path is classed as a steep hill walk, or is, in fact, a proper mountain climb, as I've not been able to identify it by name. In all honesty, it felt as though we had climbed Mount Everest!

I had had visions of us having to be plucked to safety by mountain rescue teams, and receiving a stern ticking off for taking our Scout on such a perilous journey. But if he was able to say, I'm sure that Scout would have given the whole experience a ten out of ten from start to finish!

12 A dream come true

We often daydreamed with friends about taking a holiday together with our dogs. We very much enjoyed each other's company, and liked nothing more than spending time together. Collectively, we were ten adults and thirteen dogs, perhaps a bit of a tall order when it came to it. But what are dreams for if not to turn into reality, and we were fortunate enough to do just that one autumn.

The first decision to make was: where to go? We needed to be as close together as possible; we needed lots of space; we also needed to ensure that, wherever we went, it was safe for Scout, and, above all, we needed somewhere that would welcome our dogs with open arms.

Our friends, Jo and Neta, came up with the perfect holiday destination: the tiny coastal village of Sea Palling in Norfolk. Having spent many holidays there with their dogs, they felt it would be the ideal venue.

Decision made, we booked our accommodation: a row of modern cottages just a few hundred yards from the local beach. We booked out of season, too, choosing the last week in October, because peace and solitude were what we sought, as well as a little bit of late autumn sun, if possible.

Our happy band of travellers comprised me, Paul, Jo, Neta, Nicola, Jon, Fay, Darrin, Betty, and Eluned, and the dogs: Indy, Izzy, Gracie, Buddy, Susie, Bliss, Shadow, Mila, Cindy, and Larry.

As the date of our holiday approached, our excitement grew.

I well remember arriving at our cottage that first day of our holiday (our friends had arrived a few days earlier, and were already well settled in).

As soon as we arrived, we let Misty, Sparky, and Scout jump out of the car to explore their new residence. As usual, all eyes were on Scout as he began to find his way around the secure front garden, nose and whiskers twitching in customary fashion. Once he had completed a full circuit, Scout was ready to go inside. I tapped on the front doorstep to bring it to Scout's attention, and, once his front legs connected with it, he paused, dropped his head, touched the step with his nose and hopped inside.

I followed Scout inside and watched as he created a mind-map of the internal layout of the downstairs area, moving around the rooms' periphery, negotiating furniture with complete confidence. Once he had completed downstairs, Scout tackled the steep and narrow stairs to repeat his mapping of the upstairs rooms.

I never grew tired of watching Scout perform this task, making the unfamiliar familiar; working out where obstacles were, and figuring out the best and safest way to jump up onto settees, chairs, and beds.

Once he had completed his mapping, Scout received a tasty treat as a reward for his endeavours, and we hurriedly unpacked our things, most of which belonged to the dogs, anyway: beds, blankets, toys, bowls, food, treats, and various coats and collars. Mine and Paul's luggage consisted of one tiny holdall.

Our cottage was at one end of the row, so we passed our friends' cottages en route to the beach, spotting Gracie and Buddy sitting in the downstairs front window of Jo and Neta's cottage ahead, just watching the world go by. Both dogs began to bark excitedly when they saw us, which threw Scout and Sparky into a bit of a tailspin, because they seemed to recognise the barks, but could not work out why they were hearing them in such an unfamiliar setting.

Scout began to spin on his lead, barking noisily, whilst Sparky scanned the area looking for Gracie and Buddy. Walking past Jo and Neta's cottage, Sparky cast a cursory glance toward the property, doing a double take as he realised that his friends were here, too! Capering around on the end of his lead, Sparky skipped and danced his way along the path as we continued toward the beach.

When holiday brochures describe beaches as being 'right on your doorstep' I'm always rather doubtful about just how close they are. But the truth was that the beach at Sea Palling literally *was* a stone's throw from our cottage. Accessed by a steep concrete ramp, we discovered what we fondly refer to now as a 'little bit of paradise': a vast expanse of golden, sandy beach that stretched in both directions as far as the eye could see. The tide was out that day, and small, white-tipped waves lapped gently at the shoreline. The sea was calm, due, we later discovered, to the rocky reefs that encircled the beach. The reefs had been created to form part of the local coastal defences, and, every now and again, if you were very lucky, it was possible to spot a seal or two bobbing up and down in the water in the distance.

Scout's excitement was growing by the minute as he lifted his head to take in the new smells flooding his nostrils, shuffling his paws in the soft sand. He knew he was on a beach; his all-time favourite place.

We unclipped all three from their leads, and Sparky was the first one to go haring off across the beach until he was a mere speck in the distance. Scout quickly took off in hot pursuit, honing in on the sounds made by the metal discs on Sparky's collar. Misty chose to follow at a more sedate pace, breaking into a steady trot rather than a full-on gallop.

As previously mentioned, beaches are the ideal place to give Scout the freedom to run. Flat, open spaces, with no obvious hazards or obstructions, give us the confidence to let him off his lead, safe in the knowledge that there's no danger he'll injure himself. Scout seemed to know that he was completely safe, and, with muscles rippling and legs at full stretch, our mini dynamo at full speed was an amazing sight to behold.

Sparky eventually wheeled and zoomed back toward us. Scout hadn't quite managed to catch up with him, but must have felt the rush of air as Sparky whizzed by as he came to a stop, turned around, and ran back to us, thereafter contenting himself by running around us in wide circles, barking now and again in sheer joy.

We had been on the beach for about half an hour when I happened to look back toward the ramp that served as the entrance to the beach, to see a sight that made my heart swell: all of our friends, together with their glorious collection of hounds.

We came together in a joyous tangle of leads, legs, and furiously wagging tails. What was once just a dream had become reality – what an achievement!

The rest of that first afternoon was spent strolling along the beach together, planning the week's activities. We rounded off our walk by indulging in sweet, sugary doughnuts, freshly cooked in a small coffee shop situated at the entrance to the beach, And, yes, the dogs got their fair share of the sweet treats, too.

That first evening in Sea Palling we established a new tradition, which has endured to this day: as darkness fell, we made our way over to Jo and Neta's cottage, for a delicious, home-cooked meal. Under the watchful eye of Buddy (who had assumed the role of kitchen supervisor at the head of the kitchen table), Jo – together with one or two willing volunteers – knocked together some truly mouth-watering dishes.

Once dogs and humans alike had been fed, we gathered in the spacious front room, and whiled away the hours before bedtime relaxing in each others' company, the dogs forming a canine carpet around our feet. You might think that, after such a busy day, they would all sleep soundly where they lay, and mostly they did ... except for Bliss, that is, who managed to doze with one eye open, seizing every opportunity to sneak onto one of the sofas the minute someone vacated the space. Bliss was adept at neatly folding herself into the smallest of spaces, making it look like she had been there for hours. No-one really had the heart to disturb her once she had curled herself into a tight ball in a comfy spot.

The only other disturbance was if Scout decided he wanted to move from one resting place to another. With his usual admirable skill, he mostly managed to manoeuvre himself without trampling all over his companions, but if he ever did accidentally tread on one of the dogs, it would always be Cindy, who responded by giving him a firm telling-off. Scout always took this rebuke with good grace, and then quietly sidled away.

Halloween at Holkham Hall

Having never been to Norfolk before, our holiday itinerary was put together, for the most part, by Neta and Jo, who had spent many holidays in Norfolk, and so had built up a wealth of knowledge about the local area and the best places to visit.

Our first full day out together was to Holkham Hall. Halloween was just around the corner, and Holkham Hall was holding a special event. To add a unique twist to the day we had decided to dress the dogs in Halloween costumes, and, sporting a fetching witch's hat which she wore at a jaunty angle, Mila stole the show that day, with Larry a close second in his ghoulish tutu. Misty, Sparky, and Scout wore matching striped jumpers with skull and crossbones emblazoned on the back.

Driving through the imposing iron gates at the entrance to the Holkham estate gave us our first taste of the Halloween theme, as they were fully adorned with ghostly decorations that fluttered and danced in the breeze.

> **❝ I had to blink back tears as I watched Scout happily nosing through the torn paper ... ❞**

Gathering together our motley crew of witches, wizards, pumpkins, and ghouls, we set out on a walk round the extensive grounds of the Hall. We followed a path that took us around the Hall's impressive lake, stopping now and then to admire the herds of Fallow deer that resided at Holkham. We also had to stop many times along the way to have our photographs taken by fellow walkers, delighted with our collection of dressed-up dogs.

Walking a full circle we ended up back at the Hall, where we took the opportunity to rest for a while in an adjoining courtyard, and enjoy steaming mugs of creamy hot chocolate. The dogs continued to thrill other visitors there, and obligingly posed for photographs.

I think sleep came easily to us all that night.

The Birthday Boy

The following day was Halloween, and a special day for Scout as we had chosen this date as his birthday. Of course, we didn't know when his actual birthday was, so picked a day that was as quirky and unique as Scout himself!

We got up early that morning and headed straight to the beach to give Scout, Sparky, and Misty a good run, before calling on Jo and Neta for a hearty cooked breakfast to set us up for the busy day ahead.

Scout was greeted by a rousing chorus of *Happy Birthday to you!*, and showed his appreciation by treating everyone to a generous face lick. Opening his birthday presents, Scout was ably assisted by Buddy and Susie. Eagerly, Scout tore the wrapping paper from each present, revealing squeaky toys, treats, a smart new collar and lead, and a beautiful, homemade fleece blanket. I had to blink back tears as I watched Scout happily nosing through the torn paper before choosing a ball, which he began to squeak enthusiastically. Surrounded by those who loved him most, cared deeply about him, and wanted to be a part of his life, what more could our little Birthday Boy have wished for?

The plan for the day was to take a trip to Horsey village to pick up another circular walk, that included the famous Horsey Windpump, an early 20th century, wind-powered drainage mill. The walk also afforded us some beautiful views across the Norfolk Broadlands, with its mere, marshes, and reed beds.

The weather that day was glorious: blue skies and bright sunshine accompanied us as we made our way at a steady pace, taking in all the stunning scenery. Scout loved this walk, and tripped along happily alongside us, nose in the air, sniffing out a vast array of new and exciting scents.

We finished off our afternoon outside a small café, enjoying hot buttered teacakes and a much-needed cup of tea.

After a short drive back to our cottage, we joined the others at Jo and Neta's cottage to continue Scout's birthday celebrations. And celebrate he did. Unbeknownst to myself and Paul, Jo and Nicola had prepared a special birthday cake for Scout made from all his favourite ingredients, which was presented to him along with another raucous rendition of *Happy Birthday to you*! Scout may not have been able to see the cake, but his nose told him it was something tasty, and he hopped from one foot to another in anticipation as it was placed on the floor in front of him.

Expecting Scout to have a little sniff, and then a dainty nibble at some of the treats decorating the top of the cake, I was quite unprepared for what happened next. Scout was so excited that he leaned too far forward – and fell face-first into his lovely, hand-crafted birthday cake!

Not one to let a golden opportunity pass by, Scout began to demolish the cake at lightning speed, regardless of the fact that his whole face was now completely smothered.

Left to his own devices I have no doubt that Scout would have polished off the lot, but I did manage to salvage some squashed remnants, which we divided between the hungry crowd of canine onlookers. Scout looked rather miffed when the cake was removed from under his nose, but soon cheered up when he realised that he was still wearing quite a lot of it.

And what else did we do on that first holiday?

Well, we enjoyed a stroll along the promenade at Great Yarmouth; ate fish and chips on the cliff tops at Cromer, overlooking the pier ... and even did some seal-spotting at nearby Horsey Gap, which is home to hundreds of these creatures.

Scout managed to complete a long and quite arduous coastal path walk from Horsey Gap to Sea Palling, in company with Sparky, Buddy, Susie, Mila, and Izzy; tackling some quite difficult terrain along the way – including steep sand dunes, narrow paths, and dense vegetation – but managed to keep up with his friends. Scout's reward at the end of the walk was a share in an ice-cream sundae.

But what we enjoyed more than anything else was being amongst true friends.

We promised Scout to keep a bright light burning in his dark world: what brighter light is there than that of friendship and love?

Scout and friends attending a Christmas fund raiser.

The Whippets of Oz! (Courtesy Jo Green)

e fastest recall along Temptation Alley! (Courtesy Jo Green)

Scout trying out agility. (Courtesy Jo Green)

Scout and I after being invited to judge a novelty dog show.
(Courtesy Jo Green)

A family walk with Scout taking the lead.
(Courtesy Sue Vought (www.svphotography.co.uk)

Scout demonstrating the perfect lean.
(Courtesy Jo Green)

Living the dream!
(Courtesy Jo Green)

Winter wonderland ...
(Courtesy Jo Green)

Sausage bobbing!
(Courtesy Darrin McCallum)

Perfect
communicatio
without words.
(Courtesy Jo Greer

Scout, Misty and Sparky. (Cou
Nigel Ord-Smith www.nigelordsmith.

What a smile!
(Courtesy Sue Vought www. svphotography.co.uk)

The look of love. (Courtesy Hartstone Photography)

What a stunning boy!
(Courtesy Nigel Ord-Smith
www.nigelordsmith.com)

Scout: always full of fun; always ready to play.
(Courtesy Sue Vought www.svphotography.co.uk)

Scout, proud Pets As Therapy dog.

Scout, leading blind dogs into the light.
(Courtesy Mike Sewell www.mikeysewell.com)

13 Scout's bucket list

Another promise made to Scout was to fill his life with adventure and new experiences, so I made what I refer to as a 'bucket list' for him, comprising many different objectives that range from places we want to visit, new activities we want to try, and events we want to attend.

One goal on Scout's bucket list yet to be ticked off was achieved in August 2016. Founded in 2008, Buckham Fair – held on land owned by none other than TV actor Martin Clunes and his wife, Philippa – became an annual event in Dorset, and what originally set out to be an old-fashioned fair, grew considerably into a hugely popular day out for all the family. A packed schedule included heavy horse displays, sniffer dog demonstrations, vintage cars, and – real crowd-pleasers – a horse and pony show, and a dog show with novelty and pedigree classes.

In 2012, Buckham Fair also introduced a Best Rescue/Re-homed championship. Rescue dogs could compete in qualifying heats at local dog shows in Dorset, Somerset, and Devon throughout the year. Dogs lucky enough to be placed in these heats went forward to the championship event at Buckham Fair.

On the day of the fair, each dog and handler was invited to meet a panel of judges, before taking part in a grand parade in the main ring. The winning dog would then be chosen, with awards presented by Martin Clunes. Prize money was also awarded to the charities nominated by the owner of the winning dog.

The fair promised to provide all the ingredients for a perfect day out, and we added it to our holiday diary for 2016, deciding to book a few days away somewhere nearby so that we could attend the event.

We found a studio to rent in a small town called Langport, an ancient market town boasting many historical buildings and scenic walks along the River Parrett. One of the main streets in Langport, Bow Street, is lined with some very old houses that, curiously enough, leant backwards ever so slightly. It's believed that the fronts of these buildings were built on the foundations of the original Roman Causeway; the reason why they stand at such an odd angle.

We took a slow drive to Langport and checked into our studio accommodation. Letting Scout go first into the property to mind-map his new holiday home, Paul and I followed, with Misty and Sparky bringing up the rear. The studio was quite small, comprising just two rooms: an open-plan kitchen/lounge/bedroom, and shower room separate, making it easy for Scout to chart the area. Completing two full circuits of the room, Scout jumped on the bed and settled for a nap.

Langport is such a pretty town, steeped in history, and we thoroughly enjoyed wandering the old, narrow streets before picking up a walk along the banks of the River Parrett, where Scout, Sparky, and Misty took the opportunity to blow off a little steam after their long car journey. Sparky managed to surprise a Heron who was hiding amongst reeds on the river bank, and the bird took flight, departing with a few choice words for Sparky.

Back at the studio we decided to have an early night so that we would be refreshed and energised for the fair the following day. Somehow, we all managed to squeeze onto the bed, and, despite it being a little uncomfortable, we did manage to sleep.

I woke early the next morning, mostly because both of Scout's front paws were digging into my back. Paul, on the other hand, was buried under a furry blanket made up of Sparky and Misty, who had managed to drape themselves over him during the night.

After a quick breakfast, we set off for our big day out, full of excitement!

Buckham Fair

Our journey to Buckham Fair took about an hour, and led us down some very narrow, winding roads through picturesque villages, their quaint cottages bedecked with hanging baskets bursting with summer blooms in a riot of colours.

As soon as I saw the first road signs for the fair I felt a little thrill of excitement. I had been dreaming about this day for so long, and now it seemed that this particular dream was about to become reality.

The fair was well signposted, and we found our way to the event car park with ease: within minutes we were walking through the main entrance, to be instantly bombarded by an array of sights, sounds, and smells.

The first thing we spotted was the old-fashioned fairground, its brightly-coloured helter-skelter, noisy dodgems, and a showy ferris wheel towering above us, treating its occupants to a panoramic view of the area as it slowly rotated. The tantalising smells that wafted our way from the many different food concessions were more than enough to make our mouths water.

All around us we could hear excited chatter and laughter, and, even though it was still quite early, the fair was already busy.

Scout was in his element. Nose held high, sampling all the different scents that were being carried by the breeze, he walked happily at our side. He always coped incredibly well at busy events, negotiating crowds with ease; weaving his way around people, dogs and prams with admirable skill. He never once seemed to become worried or stressed by being amongst so many people. Despite this, we kept him on a short lead, and as usual, tried to hazard-spot on his behalf.

We made our way to the registration tent for the novelty dog show, and registered Scout, Sparky, and Misty in a few classes before taking a slow walk down to the showring.

The classes began late morning, and were all quite full. We watched the Junior Handler's class, and the Best Dog and Family class; before we knew it, entrants for the Best Rescue class were invited into the ring.

This was our cue. I took a minute to smooth Scout's coat, and brush off a few bits of straw stuck to the feathers of his tail, then guided him into the ring.

It was a very busy class, and it took the judge a while to reach us.

I kept Scout occupied with treats, and engaged him in a few training exercises to help pass the time. Once the judge reached us she bent down to greet Scout, who gave a gentle wag of his tail. The judge asked me for a little of Scout's background, and the challenges we faced when first we adopted him, and I explained the difficulties involved with taking on a partially-sighted dog, without benefit of prior knowledge or experience, though added that most of the rehabilitation work had been done by Scout, who led the way.

Because the class was quite large, two judges had been employed to work their way around the ring, and, once they had completed their task, they came together to decide the overall class winner.

After some deliberation, they reached a decision and you could have knocked me down with a feather when Scout's name was called, and he was invited into the centre of the ring to receive his red rosette. Scout's ears had pricked when he heard his name called, and obligingly he trotted out, to a gentle round of applause. He'd done it, and, by winning this class, Scout qualified for the Rescue championship the following year. I felt like I was walking on air, and so proud to be handed his winner's rosette, a qualifier's rosette, and a goody bag, which Scout decided to open there and then. I pinned Scout's rosette to his lead, and walked him out of the ring back to Paul, waiting with Sparky and Misty. The beaming smile on Paul's face said it all.

I was so caught up in all the excitement that I almost forgot we were automatically through to the Best in Show event in the grand finale. As we all trotted around the inside of the ring, Scout was on top form, striding out, high-stepping all the way; his naturally flowing gait gave him such an air of confidence, as he effortlessly moved around the ring. He appeared to know when he was performing to a crowd, and always seemed able to step it up a gear.

I felt the usual flutter of anticipation as the judge prepared to announce both Best in Show and Reserve Best in Show. Tension mounted as the judge gestured to me and the owner of a very dapper-looking little Jack Russell terrier to take a couple of steps into the ring's centre. Nervously, we stood side-by-side as the judge walked toward us, bearing the winner's rosette and trophy. Incredibly, Scout had aced it again, as Best in Show in the novelty classes at the 2016 Buckham Fair was awarded to our beautiful boy.

As we stood together to have our photographs taken, I congratulated the owner of the Jack Russell for winning Reserve Best in Show. As we left the ring, I was congratulated on our success by several people, and Scout took every opportunity to charm them!

Meeting our hero
Gathering our thoughts, we rested for a while before the afternoon show classes began. These classes were very special indeed, as they were to be judged by none other than Martin Clunes. I had entered Scout in the class 'The dog the judge would like to own,' which attracted many entrants.

We managed to find a place to sit

ringside and watch the first few classes. I could hardly contain my excitement when I caught my first glimpse of Martin Clunes, looking perfectly at home in the showring, and wearing the biggest smile: it was obvious he was thoroughly enjoying himself. Surrounded by dogs of all shapes and sizes, he was in his element.

> **Scout responded straight away by smothering Martin's face with warm, wet licks ...**

Eventually, our class was called, and all of its entrants filed into the showring. By this time, the lovely blue skies we had been blessed with all morning had been chased away by ominous-looking grey clouds, and soon enough, rain began to fall in a light though persistent drizzle.

Not being over-keen on the prospect of getting wet, Scout decided to stand behind my legs to shelter, which is where he was when Martin approached. Martin stopped in front of me, said 'hello,' and checked Scout's entry number against a list he was carrying. I don't think that he noticed Scout initially, tucked away behind my legs. True to form, however, once Scout sensed that someone was standing in front of him, he moved forward until his nose bumped against Martin's shin.

Martin looked down at Scout, and said in a very quiet voice "Oh, he is beautiful," before stooping to gently stroke Scout behind his ears. I explained that Scout couldn't see too well, but now that he had fixed on where Martin was, he was unlikely to want to move in a hurry.

Scout then adopted his usual leaning stance, pressing against the side of Martin's leg. I laughed and explained that Scout was a classic leaner, using this as of way of preventing anyone leaving, which Martin seem to find very funny. He then began to play a little game with Scout, hopping backwards one step at a time which meant that Scout had to keep moving with him in order maintain his lean. It was funny and heart-warming to watch them having fun. Eventually, Martin moved on to complete the rest of the judging, and Scout returned to his original position behind my legs.

Once Martin had completed a full circuit of the showring, he stood back to look at all the dogs one last time before announcing the winners.

Scout was awarded a very commendable fourth place, and we took position in the final line-up to be awarded our rosette, and have our photograph taken. Martin stood behind me as we were snapped by the official photographer, to once again feel the now-familiar bump of Scout's nose against his leg.

"Hello, poppet," Martin said, and leant down again to stroke him. Bingo! Scout responded straight away by smothering Martin's face with warm, wet licks, to well and truly leave his mark! It was the perfect end to a perfect day.

After taking one last walk around the showground, we made our way back to our studio, tired but very, very happy.

A big tick on Scout's bucket list, and another promise kept.

14 Alfie

As well as our promises made to Scout, and whispered messages of hope intended to reassure and underpin his belief that he deserved his place in the world, I had also made a special promise to a very important person in my life, my dear dad.

Dad humoured my ongoing obsession with my four-legged family with good grace. He liked to give the impression he thought dogs were okay in general, but that he could take them or leave them. However, scratch the tough exterior, and inside he was as soft and sweet as a marshmallow. Dad really was a big softy at heart when it came to our family pets.

Sadly, dad suffered with ill health for many years, and had to take early retirement because of it. During those retirement years, he was kept company by Alfie, an irascible Cairn terrier who had been rescued by East Midlands Dog Rescue, and Winston, a cheeky Pug cross, liberated from a desperate situation at the tender age of ten weeks and taken in by East Midlands Dog Rescue (as you can tell, adopting dogs from EMDR is a bit of a family tradition). Alfie and Winston made great companions for both dad and mum, and provided them with hours of fun and entertainment.

I clearly remember the first day that dad met Scout. I had walked Scout to their bungalow to try and tire him out a little (the idea being that he might then be a little calmer when initial introductions were made). At that time, mum and dad had only

Alfie, who was getting on in years, and notorious for being more than a little grumpy at times.

> **" ... Scout did not seem in the slightest bit perturbed: he had found himself a new face to lick ... "**

Mum opened the door to us, and Scout enthusiastically bustled past her to begin his usual mind-mapping of his new surroundings. He made his way down the hall and into the lounge, where dad was sitting in his favourite reclining chair resting his leg – encased in a heavy plaster cast following recent surgery – on top of a small table. Dad leant forward and called Scout, who immediately made a beeline toward the sound of his voice. There was a sharp intake of breath from me and mum as we both realised what was going to happen ... Scout's head collided with dad's immobile leg with an audible 'thunk,' but, luckily, due to the cast's weight, dad's leg did not move, and Scout did not seem in the slightest perturbed: he had found himself a new face to lick, and he set about the task with his usual vigour.

I was desperately keen for Scout's introduction to Alfie to go well, and so, once Scout had settled, we decided to bring Alfie into the room. Ears pricked and tail up, keen to check out the new visitor, Alfie made his entrance, and, in true Scout style, as soon as he became aware that there was another dog close by, Scout bounded over to Alfie, dropped into a play bow, and let loose a volley of loud barks directly at Alfie.

It was at the point that I realised Scout was preparing to throw his paws around Alfie's neck that I intervened. Gently tapping Scout on his back, he spun round to face me, whereupon Alfie scooted past to a safe vantage point behind the sofa; all you could see of him was the tip of his inky-black nose, nestled between his two front paws as he continued to observe Scout.

Scout must have felt the air move as Alfie slipped past him, and began to search for him in earnest, nose to the ground; brow furrowed in concentration. Alfie certainly wasn't going to give the game away, and remained stationary until Scout's twitching nose connected with one of his paws, and the play bowing began again. Ever the optimist, Scout was obviously hoping that Alfie would reciprocate in kind, and they could have a grand old time together. Unfortunately, Alfie was not the friendliest of dogs, and remained resolutely ensconced behind the sofa. Within a few minutes, the sound of snoring announced that it was, in fact, game over ...

Scout and Alfie were never destined to become the best of friends, despite Scout's best efforts, although Alfie did begrudgingly accept Scout as a house guest.

As for dad, I know in my heart how proud he was of Paul and I for adopting Scout.

Love and loss

Most Saturday afternoons, my sisters, Donna and Samantha, and I would meet up at our parents' bungalow to chat about the week's events, where we had been, who we had met, etc, and dad would join in with his funny stories which would have us howling with laughter.

We had all lived under the shadow of dad's chronic ill health for many years: that constant nagging worry when a loved one suffers from long-term illness meant we had endured many a sleepless night worrying about him ... and mum. Nothing, however, prepared us for the devastating news that dad had terminal cancer, with only a few short months to live.

Cancer is cruel, ruthless, and relentless, and it doesn't care how much pain and suffering it causes. Once cancer has invited itself into your lives it will wound you, and leave a scar so deep it will never heal.

We still had our 'family Saturdays,' but now in mum and dad's bedroom so we could sit with dad and keep him company. Dad still smiled, but the familiar cheeky twinkle in his eye gradually faded as each day passed.

I talked to Dad about Paul, our dogs, and what we had been doing, and tell him of my plans for the future; the goals I had set for Scout. I knew that if he could, dad would have rolled his eyes at his crazy dog lady daughter, but with a heart full of pride for all that we had achieved.

One day I told him of my grand plan for Scout to try for his Bronze award in the UK Kennel Club's Good Citizens Award Scheme.

Launched in 1992, the scheme provided a high quality standard of training for both dogs and owners, and incorporated many different obedience disciplines. Four levels of accreditation were on offer, ranging from Puppy Foundation to a series of Bronze, Silver, and Gold awards.

Scout had already achieved his Puppy Foundation award, and it was a dream of mine to see if the Bronze award was within our grasp. And so, in the spirit of my promises to Scout, I promised my dad that we would try for the Bronze award in his honour; for what he had fought and overcome with great courage and dignity.

Dad died on 3 May, 2015, and with him went a piece of our hearts. My heart tells me he is still watching over us: the twinkle of stars takes me back to those lazy Saturday afternoons, and the twinkle in Dad's eyes which spoke volumes about the love he had for his family.

Scout goes for bronze

I found myself, the morning after dad had passed away, sitting alone with Scout at my side, contemplating attempting the bronze award in dad's memory, trying to pretend that everything was normal.

I had suggested to Paul we could attend a local event that was running the Good Citizens awards as part of the schedule. I think he thought I had taken leave of my senses, and urged me to reconsider, and maybe take the assessment at a later date. But determination had set steel into my spine: this was for dad; my mind was made up; we were going to give it a try!

I sat there for maybe an hour or two, lost in thought. Now and again my musing was interrupted by Scout, who nudged me to check for fresh tears, which he dutifully licked away.

Eventually, my reverie was broken as our names were called. I stood up, took a deep breath, and tried to keep my hands from shaking as we walked toward the fenced enclosure where the majority of the assessment would be carried out.

Our examiner for the day put me at ease straight away as she greeted us with a warm smile, before stooping to say hello to Scout, giving him a reassuring ear rub as she did so.

I briefly explained about Scout's sight problems. The one aspect of the assessment that did concern me was being able to demonstrate Scout's ability to return to me when called whilst off-lead, as a reliable recall was one of the listed criteria for the Bronze Award.

Scout's recall is impressive, as previously mentioned; he would always head for the sound of our voices, and, if it was windy and our voices carried, would recall to the spot where he *thought* the sound was coming from. I didn't think that the slight breeze that day would be too much of an issue for us, as a result. The examiner assured me that she would proceed slowly and steadily to allow Scout to take things at his own pace.

To begin the assessment, we had to pass through a little gate into the enclosure, to demonstrate our ability to do this in a calm and orderly way. I asked Scout to wait whilst I undid the catch on the gate (my hands were trembling, and I fumbled with the latch a couple of times before finally lifting it). Opening the gate fully so that Scout didn't bump into it on his way through, I passed through, turned to face him, and called him through. Right on cue Scout walked straight through the gate to take up his usual place at my side.

Next, the examiner asked me to walk Scout around the inside of the enclosure on a slack lead, and we did a couple of full circuits that included stops and turns. For sudden turns, I would try to indicate to Scout that I was about to change direction by allowing him to make contact with the side of my leg, in order to steer him into a different direction. On a tighter lead I would have achieved this by applying gentle pressure on the lead to encourage him to turn left or right, but, as the requirement for the assessment was a slack lead, I used the side of my leg to guide him. We did several turns which Scout executed perfectly, before returning to the examiner.

I was then asked to remove Scout's collar and lead. The examiner checked the identity tag to confirm it had the correct information, before returning it to me to put back on; she also checked that I was carrying a supply of waste bags (I had a large quantity crammed into the back pocket of my trousers).

The next part of the assessment entailed demonstrating that Scout was happy to be examined from nose to tail, and was also comfortable with being groomed. This part of the exercise posed no problems at all as Scout loved to be touched and groomed.

I started at the front, and lifted Scout's ears one at a time. Next I lifted his lips and examined his teeth, ran my hands down and over his back and tummy, and picked up his paws, one at a time, to inspect his nails. Scout thought this was great fun, and started to try to lick my face as I knelt in front of him. Grooming a short-coated dog such as Scout was relatively easy, especially with the

grooming mitt I used to keep his coat silky-smooth and dirt-free.

Whilst Scout enjoyed his mini pampering session the examiner asked me a few general questions about a dog's needs, and my responsibilities as a dog owner. I was fairly confident in the answers I gave for each of the questions asked.

> **❝ ... I felt the tears pricking at the corners of my eyes when he eventually broke into a gallop, mouth wide, tongue lolling ... ❞**

Scout was now totally relaxed, and seemed to be really enjoying himself, so we continued on to the next part of the assessment: the requirement for him to remain in one place for exactly a minute whilst I moved a short distance away. Scout could adopt any position he wanted for this – stand, sit or down – as long as he maintained that position for the time required. The exercise would be complete when I returned to Scout's side and picked up his lead.

I had practiced this time and time again with Scout, and he was reasonably comfortable with being asked to perform a 'down-stay.' The problem was that, if I used my voice to encourage Scout to remain in the down position, he tended to move, as he wanted to be where the sound of my voice was coming from. To try and prevent this, I would have to remain silent throughout the time period, which could make Scout curious about why he couldn't hear my voice, potentially causing him to break the down-stay to try and locate me.

I gave Scout a single loud 'down' cue, and, after thinking about it for a few seconds, he obliged. Without further ado, I dropped the lead next to him, and quietly took a few paces backwards. The examiner then began to time the minute.

A minute can seem like an eternity. I watched Scout as he started to wriggle a little, and, although once or twice it did look as though he was going to stand up, he held his position for the full minute, and was rewarded with a big hug when I walked back to him and picked up his lead.

Thank you, dad

Next came the part of the assessment I had been dreading – the recall. I knew that Scout would respond to me when called, but just how far off target might he end up? The other complicating factor was that the dog would be encouraged to engage in a form of play away from his or her owner, in order to interrupt his or her focus on the handler, but Scout always had to be close to us to do this as he couldn't find a toy if it was thrown for him. Scout loved to play with other dogs, and would happily give chase if a ball was thrown for another dog, but he wouldn't go after it himself. Likewise, Scout loved a good game of tug, but would never initiate it.

I explained this to the assessor, who said that as long as Scout was a reasonable distance away from me when he was recalled, it would be sufficient, as the fact that he would be left alone for a short time would provide adequate distraction. TFeeling a little apprehensive, I walked Scout to one end of the enclosure, unclipped his lead, and asked him to wait. I then made my way across to the other side of the enclosure, and turned to face him. Scout stood perfectly still, nose in the air, as if trying to pick up my scent. Once the assessor was happy that I was a reasonable distance from Scout, she asked me to call him.

In my most high-pitched, excitable voice I called Scout's name, then crouched low to the ground, arms outstretched, and repeated his name loudly and clearly. The second that he heard my voice, Scout's ears pricked, and he began high-stepping his way forward, slowly, at first, as he was on unfamiliar ground, and also had neither dog nor person at his side to bolster his confidence. Scout was doing this all by himself. I felt the tears pricking at the corners of my eyes when he eventually broke into a gallop, mouth wide, tongue lolling in a beatific smile.

I caught Scout in my arms and buried my face in the warmth of the soft ruff of fur around his neck, drying the tears on my cheeks.

All that was left to do now was the last part of the assessment: demonstrate Scout's ability to walk calmly amongst people and other dogs.

By now, it was early afternoon, and the venue was getting quite busy. The main event of the day was a charity dog show, which had attracted a large number of people and their dogs, as well as several stalls and dog-related activities: all-in-all Scout had a lot of distractions to contend with. Undeterred by this – and buoyed by doing so well in previous exercises – we were raring to go!

With the assessor at our side, we left the enclosed area via the gate. As before, I asked Scout to wait whilst

I unlatched the gate (with greater success this time as my hands had finally stopped trembling), and began to make our way toward the nearest gathering of people and dogs. As we walked, I talked to Scout in a low, quiet voice, reminding him I was with him, and there was nothing to worry about.

We reached the first group of people and dogs, and successfully threaded our way through. One or two dogs reached out to sniff Scout as we passed, but he remained glued to my side, concentrating totally on the sound of my voice.

Next, we walked behind a row of stalls where there were fewer people and dogs. Taking care not to let Scout to trip over the guide ropes securing the stalls, we picked our way around, passing people and dogs without reaction from Scout.

The assessor then asked me to stop at the next person we came across, and briefly engage them in conversation, with Scout remaining calm and quiet whilst I did this. I approached a woman standing nearby to ask if she would mind helping us with this part of the assessment, and she was more than happy to do so. We stood chatting for a few minutes as I explained to her that we were being assessed for an award, and ran through what we had done thus far, with the assessor standing to one side throughout, observing Scout's behaviour. After giving the woman's leg a perfunctory nose bump, Scout was happy to stand and listen in on our conversation. Ending our conversation, I thanked the woman for her time. She wished us luck, and was rewarded with one of Scout's extra special leans.

And that was it! We had succesfully completed all of the individual assessment components, and our fate rested in our assessor's hands. Had we managed to make the grade?

Wearing the biggest smile, our assessor congratulated us, and confirmed we had achieved our Bronze Good Citizens award!

I looked skyward and tipped the heavens a wink. I had felt my dad's presence throughout the day, and knew he was smiling down at me. Dad had given me the strength and confidence to guide Scout, and between us we had achieved our goal.

Several days later I noticed a post on social media written by our assessor, thanking all those who had attended that day, and congratulating them on their achievement. She wrote that the highlight of her day was helping the owner of a certain short-sighted dog secure the bronze award, and commented on the strong bond that clearly existed between dog and handler.

I was so touched by her words that I dropped her a message of thanks, explaining the circumstances which had led me to try for the award.

Scout has been and continues to be instrumental in helping me deal with the loss of my dad, and I have both good and bad days in this respect. On the good days, he will raise a smile with his clownish antics and tomfoolery, and on the bad days he will lie patiently and sleep at my side for hours. Scout will lick away my tears, and is always on hand for a comforting hug. Inasmuch as I have always been there for Scout, he, in turn, has led me through my mourning.

15 The social touch

Being a bit of a social butterfly means I am always happy to step up to the plate with Scout, and help educate the public about how life-enhancing a disabled dog can be. Scout's ability to charm, coupled with his striking looks, makes him a head-turner, and rarely does he ever fail to draw a crowd.

This ability also provides us with many opportunities to fulfil another promise that we made to Scout: to try and give back to those who had helped in his time of need. And what better way to repay the kindness shown than by getting out there and helping to raise awareness, and much-needed funds, for our local charity, the East Midlands Dog Rescue?

Over the years, we organised (with a lot of help from friends and family) and attended many fund raisers. Our bi-annual tombola market stalls – run with friends in spring and at Christmas – were always popular. Armed with boxes and boxes of donated prizes, we met up with fellow volunteers at first light to set up in our town centre, and learned a lot about how to run a successful tombola stall from our early experiences, such as putting out the prizes in numerical order rather than untidy piles; remembering to keep the lid on the tombola ticket box, especially on windy days, and always taking a spare pair of woolly socks to warm our frozen feet. Everyone pitched in and helped sell tickets, give out prizes, and rattle collecting tins, and, together, we laughed, smiled, and joked our way through the day.

What made the biggest difference to our fund raising totals was taking along our dogs, all of whom behaved impeccably. And so to Barney, Floyd, Mac, Hero, Gabriel, Teddy, Ramona, Brooke, Hamish, and not forgetting our very own Izzy, the fundraising hounds who are no longer with us, I dedicate this chapter.

Scout loved these events, and got stuck in from the moment he arrived, greeting every visitor to our stall with a tail wag; giving out endless cuddles, and enthusiastically licking any face offered to him. What made him the perfect candidate for this type of event was his innate ability to adjust his greeting to suit. With young, exuberant people, he would be the class clown, capering about, and generally acting the fool. For quieter, more reserved individuals, he became quieter himself, opting to lean against their legs to allow them to gently stroke him.

Visitors to the stall were always genuinely keen to hear about our dogs, and their stories, just like Scout, many of the dogs who attended these events had triumphed over adversity – survivors, one and all. This also provided the ideal opportunity for questions about rescue dogs, which we were always happy to answer in the hope of dispelling some of the myths that exist about them. Some dogs end up in rescue through no fault of their own: people's circumstances change – relationships fail, health problems occur, and jobs are lost – life-changing events beyond their control, which sometimes mean that relinquishing a much-loved animal companion is the end result.

Other dogs are passed to rescue because their owners realise – too late – that owning a dog is a huge commitment, both emotionally and financially. Researching the breed that best suits a lifestyle, putting time and effort into training, and properly understanding a dog's needs are all essential requirements to taking on a dog.

Far too many dogs are signed over to rescue with problems that could so easily have been avoided if the right decisions had been made from the start.

Many of the fundraising dogs were rescue Greyhounds, so these events provided a perfect opportunity to educate and spread the word that these fabulous dogs make great pets.

The Greyhound racing industry operates worldwide, and, every year, thousands of Greyhounds find themselves surplus to requirements once their racing careers have ended. The lucky ones end up in rescue; others face a less certain future, and many will never find their forever home.

If you've never talked with the owner of a rescued ex-racing Greyhound, you're missing out on a real treat. The love and respect they have for this elegant, graceful breed is unmatched, and is often evidenced by the fact that they rarely ever have just the one animal.

Greyhounds are still very much misunderstood as a breed. Despite their size and strength, they tend to be very steady dogs to walk on the lead, and are quite happy to trot companionably at your side. Greyhounds are the ultimate couch potatoes, and like nothing more than a good snooze on a comfy bed. As with any other breed, they do require exercise but, as a rule, most will be more than content with two twenty minute walks a day. Greyhounds have been known to live happily alongside other small breeds of dog; cats, rabbits, and even chickens, although this does depend very much on the individual dog. Greyhound-specific rescues, and those with experience of Sighthounds, can offer a great deal of help and support to those considering adoption. No two Greyhounds are the same, of course, but what you get with one – true love and devotion – will always be the same. Scout's Greyhound friends, Shadow and Larry, are living proof of this.

Overcoming challenges

As well as organising and taking part in fundraising activities to help raise awareness of dogs in rescue, we were also keen to demonstrate Scout's ability to undertake a variety of new challenges. In this endeavour, we have had to make a few tweaks here and there, and have sometimes had to recruit the help of complete strangers to facilitate Scout's needs, but have discovered that people are only too pleased to lend a hand to enable Scout to try his paw at new activities: what I fondly refer to as the 'Scout effect.'

The many canine-themed events we attended featured many things we wanted to have a go at with Scout, and fun agility was one I was keen to try. I loved watching the dogs as they flew around the agility courses, jumping through suspended hoops, scaling A-frames, and dashing through tunnels: they always looked like they were having so much fun. I wondered whether Scout could cope with a gentle trip around a course, guided by me and a volunteer agility instructor.

> **“ ... I asked him to jump, and was a little surprised when he did just that! ”**

I decided to take the plunge with Scout at a show we attended in the summer of 2016. I had a quiet word with one of the agility instructors to see if she would help us negotiate the course, and she was more than happy to do so. After showing me around the course and the individual obstacles, we were ready to go.

Keeping Scout on a short lead, we started with the high hurdles (in Scout's case the fairly low hurdles, as the bars were lowered for him). Using a treat to guide him forward, I brought Scout as close as I could to the bar of the first hurdle, which was about a foot off the ground. Once Scout's legs made contact with the bar, I asked him to jump, and was a little surprised when he did just that! Over he popped, clearing the bar with an inch or two to spare.

Feeling encouraged by this success, we moved on to the weave poles, a series of six or so in-line, upright poles; the idea being that the dog weaves their way in and out of them. Scout touched the first pole with his nose, and I then, very slowly, guided him in and out around each one, continuing to use a treat to entice him forward, and praising him as he moved with considerable dexterity around each pole.

The next challenge was an A-frame: a tall, wooden structure in the shape of an upside-down 'V,' the sides of which sloped steeply; I wondered what Scout would make of it. Once again, following the treat held at the tip of his nose, and with gentle words of encouragement, Scout began to climb, lifting and placing each paw carefully and methodically as he went, all the way to the top before slowly making his way down the other side. I've never doubted Scout's confidence, but even I was amazed by how well he was tackling the course!

Scout easily coped with a couple more jumps before we approached the next tricky obstacle, quite challenging for Scout, despite its simple appearance of a long, narrow plank of wood with a short ramp each end. The object of the exercise was to walk up the ramp, along the plank of wood, and back down the ramp the other end. Walking Scout in a straight line isn't the easiest of tasks at the best of times, so to ensure he stayed safe I walked on one side of him with the instructor on the other. I had my arm over Scout's back, with the flat of my hand against his chest to help guide him: once again, Scout confounded expectations by calmly and slowly making his way across the plank, and down the ramp at the far end without faltering once.

By this time a small group of onlookers had gathered to marvel and smile at the sight of this little dog, high-stepping his way around the course, nose stretched forward, whiskers twitching constantly, and pausing now and then to crunch one of his tasty rewards.

The next obstacle to challenge Scout was a pipe tunnel – essentially, a rigid length of corrugated material. I did feel a little apprehensive about this one, as Scout would have to enter and move through the tunnel by himself – and off-lead – without the reassurance of having me close at hand. The instructor stood with Scout at the entrance to the tunnel, and I positioned myself at the exit, kneeling with arms outstretched.

The instructor unclipped Scout's lead and edged him toward the tunnel entrance. Without further ado, Scout vanished inside. I strained my ears to catch the sound of his steady footfalls, and, after what seemed like an eternity, I spotted Scout's pink nose as he emerged from the tunnel and came straight into my open arms. I think he was glad to be reunited as he placed his paws on my shoulders, and smothered my face in warm, wet licks.

To complete the course Scout jumped another couple of hurdles with considerable style to a round of applause from onlookers. I thanked the instructor for her kindness and patience, and we made our way out of the ring, Scout happily bouncing along at my side. Did he enjoy himself? I think his wagging tail and joyful demeanour said it all ...

Run, Scout, run!

Another popular fun event at dog shows was the fastest recall competition. This fast and furious challenge was generally set as a flat run, in which the handler would stand at one end of an enclosed strip of ground whilst their dog was walked to the other end by a volunteer handler. When instructed to do so, the dog was released from their lead, and timed as he or she dashed back to their owner.

> **“... Scout always became so excited when he heard other dogs taking part in the run: barking, spinning, and trying to pull toward where the activity was taking place ... ”**

At some events, to make things a little more fun and challenging, tempting distractions such as treats, toys, and balls are scattered along the length of the run, and it's fascinating to see which dogs completely ignore the distractions, and which dogs casually stroll the course, stopping at every temptation, snaffling every available morsel of food, and pausing to pick up each squeaky toy. Izzy, Misty, and Sparky loved these recall competitions, and, between them, won many a rosette for their speedy efforts.

I had never considered letting Scout try a run, as I thought it a big ask for him to run solo on such a narrow track: my greatest fear was that he would veer off as he ran, collide with the fencing, and injure himself. But – it looked for all the world like he really wanted to join in.

It was a risk, but the question was, should we take that risk ...?

Our friends, who were as devoted to Scout as were Paul and I, had witnessed his enthusiasm and excitement whenever he was near a fastest recall track, and they managed to convince us that we had to let Scout have a go. Scout's life was all about pushing boundaries and achieving goals; it was time to see just what he could do.

We were attending a fun dog show one summer that was running a fastest recall competition, and I decided that this was the time to take the plunge. I asked the organisers of the competition whether Scout could give the run a try, and they were more than happy for him to do so. Firstly, I checked out the actual run to ensure that the ground was completely flat, and found it was full of distractions – squeaky toys, balls, and paper plates laden with treats. Paul walked Scout to the start of the run, and I made my way to the finish line. Kneeling on the ground, I opened my arms wide, which indicated to Paul he should release Scout. Once I saw that Paul had unclipped Scout from his lead, I began to call Scout's name loudly and repeatedly. Scout stood for a second or two, tilting his head at the sound of my voice calling him, and that was all it took. Without further ado, he broke into a run, paws and legs flailing, and gambolled his way down the length of the run. Rather than stopping at the temptations

that lined his route, Scout trampled straight over them, sending treats and toys flying in all directions. He was managing to more or less maintain a straight line, but I could see that, as he drew closer, he was beginning to veer to one side slightly. I adapted my position a little and continued to call his name.

Applying the brakes isn't one of Scout's many talents, and, as he came thundering towards me, I realised that he wasn't slowing. I closed my eyes and braced myself as Scout crashed into my open arms.

Being hit by a speeding dog is not something that I would recommend. Scout's solid frame connected firmly with my chest, and he sent me sprawling. Opening one eye, I saw blue sky above me interspersed with little powder puff clouds drifting by lazily. This vision of peace and serenity was suddenly obscured by Scout as he leant over me, and began to wash my face in earnest. I lay there for a minute or two while Scout administered his own version of first aid, before scrambling back onto my feet.

Had we made the right decision by letting Scout have a go? The wagging tail and bouncing step as we made our way out of the enclosure gave me my answer.

At the end of the afternoon, the winner of the fastest recall class was announced, and we were all delighted to hear that the little dynamo Susie had clinched the top spot, and awarded a stunning rosette.

The organisers of the recall class added that they had a special announcement to make, and explained they had hoped that a dog such as Scout would enter the class that day, to demonstrate how dogs with disabilities can take part in and enjoy the same type of activities that an able-bodied dog can. Scout, Paul, and I were called to accept a special certificate, and, much to our surprise, a beautiful, bronze-coloured statue of a sitting dog, almost twice the size of Scout, which has pride of place in our front room. Once again Scout's courage and willingness to participate had shone through.

And, yes, before you ask, I did cry.

With a little help from his friends

As well as the more challenging activities such as agility and fastest recall, there were other, less demanding events that, with a tweak here and there, Scout could happily participate in, along with his friends, and, I guess if Scout could choose his favourite fun activity, sausage bobbing would likely be top of the list.

Sausage bobbing has grown in popularity over the last few years, and is a fun and engaging way to help raise funds for charity. Based on the same principle as apple bobbing, segments of sliced hot dog sausage are placed in a container of water, into which dogs dip their noses in an effort to pick out a tasty morsel or two. It's highly entertaining to watch the dogs as they try to grab a piece of sausage from the water. The true professionals were adept at flicking a piece right out of the water, gobbling it down with barely a drop of water on their whiskers. The novices plunge their noses right under the water, coming up for air now and again, and showering onlookers with water as they shake themselves dry.

Scout's friend, Buddy, took the challenge to a different level altogether by climbing *inside* the container of water, and splashing about, scooping up bits of sausage at lightning speed. You could see the joy on his face as he wallowed in the briny, sausage-filled water.

> " ... I wondered if Scout could manage to pick up a piece of sausage by using scent alone, I guess there was only one way to find out ... "

I was watching a group of dogs pit their wits against the floating sausage at a local dog show one day, and could see that the dogs were making full use of their excellent sense of smell to help locate the sausage, but I also noticed that they were using sight to more accurately pinpoint them as they bobbed up and down in the water. I wondered if Scout could manage to pick up a piece of sausage by using scent alone, I guess there was only one way to find out ...

I walked Scout to the container holding the water and sausage pieces. His nose and whiskers were already twitching as he quickly detected the sausage scent. I took him to the edge of the container of water, and tapped gently on the side of it, and Scout instantly dipped his nose into the container. I think he was a little surprised when he made contact with the water, as he lifted his head and violently snorted, spraying droplets of water, and creating a fine mist around him.

Scout knew there was something worthwhile to be had in the water, but couldn't quite work out what it was ... or how to get it. I could only surmise that his sense of smell was being overloaded because the strong scent of hot dog sausage was preventing him from honing in on an individual piece.

Always with Scout, I never set him up to fail, but the answer to this little conundrum didn't come from me, but from Eluned, one of our friends, who had been watching Scout, willing him to succeed. In an instant, Eluned scooped a few pieces of sausage from the water and let Scout sniff and then take one from the palm of her hand, which Scout did with his customary relish. Eluned let Scout continue to sniff at her hand, then gently tossed a piece of the meat into the water, where it landed with a splash. Scout responded to the noise almost instantly, turning to plunge his nose into the water; soon resurfacing with his prize. From then on Scout took to the challenge like a duck to water (if you'll pardon the pun): he simply waited for the sound of a splash and then dived straight in.

To every challenge, there is a solution; to every locked door, there is a key. For Scout, it was empathy and a determination to help our brave little hound that ensured he overcame every challenge he faced.

Visit Hubble and Hattie on the web: www.hubbleandhattie.com
hubbleandhattie.blogspot.co.uk
• Details of all books • Special offers • Newsletter • New book news

16 Inspiration

When people ask me what attributes are needed for taking on a blind or visually-impaired dog, I run through a list of qualities I like to think that Paul and I have, with empathy, understanding, and patience top of the list. I add that a sense of humour and a positive attitude are especially helpful, as the inherent worry and responsibility that comes with owning a disabled dog means that looking on the bright side of life can help you through the darker days.

One of the reasons for sharing Scout's life story is to inspire others to consider doing something similar. My inspiration comes from a little Romanian rescue dog called Nym: ex-street dog who, in an act of dreadful cruelty, had both eyes removed. A kind couple tried their best to help Nym by feeding her and keeping her safe until, one day, Nym was captured by a Romanian Dog Catcher. Nym's plight came to the attention of Blind Dog Rescue UK, a small charity dedicated to the rescue of blind and partially-sighted dogs in the UK and abroad.

Nym was brought to the UK in 2013, and began a new life with her owner, Sian, who regularly fostered for Blind Dog Rescue UK. Initially, Nym was very introverted, and terrified of everything. Sian worked hard with Nym to help restore her faith in people, and the results have been little short of miraculous, as she is now a confident, happy little dog, and a beacon of light for blind and visually-impaired dogs everywhere. Now a registered Pets As Therapy dog, Nym is a regular visitor at a residential home for children with disabilities, and is dearly loved by all who meet her.

Paul and I belong to internet support groups and forums that help bring together owners of visually-impaired dogs such as Scout and Nym. The groups provide much help and advice to people who already own a blind dog, or those contemplating adopting one. Scout was born with his sight problems, but for an owner of a dog who suddenly loses his or her sight, or has a medical condition that will ultimately lead to blindness, these groups can be a real lifeline.

> ❝ ... watching Scout's every move to ensure he doesn't bump into or trip over anything is second nature to us ... ❞

The mistaken belief that a blind dog cannot have a good quality of life still prevails, unfortunately, but, by sharing inspirational stories and images, training tips and general advice, this way of thinking can be shown as the fallacy that it is.

Getting it right

Owning a partially-sighted dog means we have got it wrong a few times, but it's reassuring to know that we're not the only ones. These are some of the daft things we find ourselves doing –

• We wave our arms around wildly whenever we do a recall with Scout.

• We shout 'look out!' when he is about to tread in something unspeakable.

• We point to things that Scout drops on the floor and ask him to fetch them.

• We sometimes forget to stop saying 'step' at the top of a flight of stairs, and Scout will continue to try to climb.

• We make the mistake of leaving trays of paint on the floor when we are decorating.

Constantly hazard-spotting when walking Scout is a skill that Paul and I have perfected over the years. Walking with our heads down, watching Scout's every move to ensure he doesn't bump into or trip over anything, is second nature to us. Adjusting our walking speed to let Scout know we are approaching potential hazards, applying tension to his lead to avoid a hazard, or using the side of our leg to keep him moving in the right direction are similarly normal behaviours. All of these tactics, although sometimes very subtle, ensure that Scout remains safe.

The importance of this essential part of our daily life was brought home to me one day when I handed over our special boy to a friend, Linda, who desperately wanted to introduce Scout to some of her work colleagues.

We had called into Linda's workplace, an open-plan office full of desks aligned in neat rows with wide spaces in-between each row. Linda came rushing over as soon as she spotted Scout and gave him a big welcoming hug, and Scout returned the favour by clamping his front paws on each of Linda's shoulders, and covering her face in licks. I could see people craning their necks to get a look at Scout, the little dog they had heard so much about.

Linda asked if she could take Scout for a little walk around the office to say hello to some of her friends, so I handed Scout's lead to her. Linda beamed with pride as she walked away, Scout tripping along merrily at her side. I guess I should have foreseen what happened next: poor Scout walked head-first into the first desk they approached.

Poor Linda was mortified; Scout was a little dazed, but otherwise just nonplussed. Even a bang on the head didn't dampen his spirits, however, and his tail carried on wagging. Realising I should have taken a little time to explain to Linda the finer points of successfully walking a dog like Scout, I caught up with them and apologised to Linda for my shortcoming.

We conducted the rest of the office tour together: Linda holding Scout's lead with me at her side, giving direction. By the time we were halfway round the office, I could see that Linda's confidence was growing. And as for Scout? Never one to hold back, he greeted everyone as if they were old friends.

17 Accidents will happen

Many of the facets of Scout's quirky way of life are beyond the scope of this book, but it would be good to touch on one or two that provide insight into how we try and manage Scout on a day-to-day basis.

Unfortunately, over the years, Scout has sustained his fair share of scratches and scrapes from bumping into things. His occasional bouts of giddiness and excitement sometimes resulted in his forgetting that he couldn't see so well, and, during some of these episodes, Scout has twirled off pavements, spun into lampposts, and collided with parked cars. Undaunted, Scout simply regained his feet, gave himself a good shake, and carried on about his business with his usual jaunty air.

The worst injury that Scout has suffered was from inadvertently falling into a tiny council inspection hole, hidden under a pile of discarded grass cuttings. Usually on the ball at spotting holes and hazards on Scout's behalf, on this occasion we weren't, as the hole was totally obscured. How Scout came to end up with both of his slender front legs wedged in a space just five inches square I will never know, but there was an audible *thud* as he fell, hitting his chin on the pavement in front of him.

Initially, I could not work out what had happened. Scout's back end was still upright where it should be, but his front end was flat to the ground. Remaining in this rather peculiar position, Scout began to whimper.

I knelt next to my dog, and gently eased his legs out of the hole, talking to him the whole time, reassuring him that he was all right, and was safe. Scout's front legs were badly grazed, and had begun to bleed in several places. His chin was also badly scraped, and tiny beads of blood had started to well, staining his usually spotless white fur. Scout began repeatedly to lick his lips, so I checked inside his mouth, and noticed that the gum above his front teeth was raw and angry-looking, suggesting that the front of his mouth had borne the brunt of the fall.

I could have cried for him as he stood there, head down, tail firmly clamped between his legs, looking sad and dejected: he suddenly seemed so vulnerable. I had never seen Scout like this before, and it broke my heart to see our brave, fearless boy looking so defeated.

To further add to our woes, Scout refused to budge, probably because he had lost confidence, and believed that the area around him was unsafe. Even reassurance from Misty and Sparky, by way of gentle snuffles and nose bumps, couldn't persuade him to take a step. I had no choice ... scooping him up and holding him tightly, I carried him home, very glad to feel his body begin to relax against me as I held him close.

Once home, I cleaned and

bathed Scout's wounds, which seemed to cheer him up immensely. I was greatly relieved to see his tail begin swishing again, and even more so when he licked my face as I checked him over.

Taking full advantage of the situation, Scout languished in his bed for the rest of the afternoon, accepting hand-delivered treats in large quantities (to 'counteract the shock,' as I later explained to Paul, when he returned from work to find Scout buried under a pile of soft blankets, crunching away on a biscuit).

Luckily, Scout did not suffer any long-term effects from his particularly unpleasant tumble, and it made us even more vigilant when out walking with him.

The only other concerning health issue we had with Scout was when he suffered from a bad bout of tummy ache.

Initially, we did not know what was bothering Scout. Sitting watching television one evening, Scout was snoozing at Paul's side when he woke up suddenly, hopped down from the settee, and began to wander in a distracted way. Worryingly, he began to bump into objects usually familiar to him, which he would normally negotiate with ease. Scout walked into a coffee table; into the side of the sofa, and then straight into a wall as he tried to make his way into the kitchen. Once there (banging his head on the door frame as he entered), Scout attempted to reach the back door, but, instead, walked straight into the side of a cupboard. As he stumbled around, Scout began to whine pitifully.

Alarmingly, it appeared that Scout had suddenly gone completely blind.

In a complete panic, I contacted our on-call duty vet and described Scout's symptoms. Andy, the duty vet, advised us to bring Scout straight to him.

❝ ... Scout languished in his bed for the rest of the afternoon, accepting hand-delivered treats in large quantities ... ❞

I don't remember much about the journey to the surgery, other than it seemed to go on for an eternity. Scout was curled up on my lap swaddled in a blanket, and now and again would groan and whine as if in a great deal of pain. I gently massaged Scout's ears to try to soothe him, telling him that help was on its way.

Once at the vet, Paul lifted Scout off my lap and onto the ground, where, a little unsteadily, he stood for a few seconds, before passing a huge pool of watery diarrhoea.

Slowly and steadily we made our way inside. Scout was still very disorientated as we guided him through the main doors, and into one of the consulting rooms. Scout leaned against the side of my leg as we walked, affording him some stability, whilst I steadied him with my hands on each side of him.

Andy was there, waiting for us, and, with calm efficiency, he lifted Scout onto the examination table and checked him over. Scout let out a loud yelp as soon as Andy touched his abdomen.

Andy's diagnosis was that Scout was suffering from a tummy bug that was likely causing him painful stomach cramps.

What Andy didn't quite understand, though, was why this appeared to have affected Scout's already limited vision, rendering him completely sightless. Gently, Andy lifted Scout off the table and onto the floor to see what he did. Scout immediately began to turn in circles. After circling several times, he then walked forward and hit his head on the side of a waste bin. Our boy was totally blind, and becoming increasingly distressed by this sudden change in his circumstances.

Admitting Scout to the hospital overnight to administer treatment and keep him under observation was the ideal scenario, Andy said, but he was concerned about how Scout was behaving, and felt that hospitalising him was likely to increase Scout's already obviously sky-high stress levels.

What Scout needed most was to rest, free from the stomach pains and distress of losing his sight completely, and Andy suggested he heavily sedate Scout to allow him to sleep whilst his body recovered. The best place for Scout was at home with me and Paul, he said, and we were quick to agree.

Within minutes, Andy had inserted an intravenous catheter into Scout's front leg, and slowly injected a powerful sedative together with a strong painkiller. I felt Scout gradually begin to relax in my arms as the sedative took effect. By the time Andy withdrew the needle from the catheter, Scout was asleep.

After one more injection to help to settle his stomach, we wrapped Scout in his blanket and took him home.

Andy was very honest with us, admitting he wasn't sure why Scout had suddenly lost his sight, although we should understand that the little vision he did have previously may not return.

Mindful of the words of my wise veterinary colleague when we adopted Scout – 'his hard wiring may not be right' – I feared the worst. Maybe Scout's external genetic defects were now beginning to have an effect internally as well. I couldn't even begin to describe the thoughts going round in my head as we made the return journey.

Once home, we carried Scout – still sound asleep – inside. It was now past midnight, and we were exhausted. We tried to decide how best to keep an eye on Scout over the next few hours. It was clear he would need to be monitored closely in case he woke up and began to panic again, so we decided that one of us would stay downstairs with him whilst the other tried to grab a couple of hours' sleep. In the end, neither of us could bear to leave him, and Scout spent the night in our bed with us.

And there he stayed until morning, our very own sleeping beauty. Other than the occasional snore Scout was as quiet as a mouse, laying between us, motionless apart from the rhythmic rise and fall of his chest that lulled us to sleep, my hand on Scout's chest, drawing comfort from the silky softness of his fur under my fingertips.

Starting over

We woke to the sound of our alarm clock, and, looking over to Scout. I could see his ears twitching in response to the shrill beep of the alarm. Gradually, Scout lifted his head, leaned across and began to lick my face. All the tension and worry of the previous night began to recede: our boy was feeling better for sure. Our concern now was his sight: was Scout still completely blind?

Slowly and still a little unsteadily, Scout got to his feet and shook himself. He took a few faltering steps toward the edge of the bed ... and then stopped. Usually, he didn't hesitate like this before jumping down, but it was clear he wasn't quite so sure in this instance. Slowly extending both of his paws, he located the edge of the bed before gradually sliding off. Once on the floor, Scout stood still, lifted his nose in the air, and began to scent, as if trying to re-establish his bearings.

Scout moved off cautiously at first, lifting and placing each paw slowly and deliberately, feeling his way. I followed closely behind, softly speaking his name over and over so that he knew I was close by. Carefully, he made his way out of the bedroom and toward the stairs. I gently rested my hand on the top of his head, continuing to speak gentle words of reassurance. It was obvious that Scout was still struggling with his vision.

Scout wanted to go downstairs, so I sat down at the top, with him on my right side. Step-by-step I 'bumped' down the stairs, patting each step in turn to encourage Scout to proceed. Each time I patted one of the stairs I paired it with the cue 'step,' just as we had always done, my free hand resting on his side, reassuring him that I was there and he was safe. One by one we tackled the stairs slowly and carefully. I heaved a big sigh of relief when we reached the bottom, and gave Scout a big cuddle. I felt Scout's tail thump against my ribs in response to the feeling of being held. 'That's my boy,' I thought, as I carefully guided him outside and onto the garden.

Scout was obviously feeling much better in himself, and ate his breakfast with gusto: the confusion and anxiety he had experienced the previous evening had all but gone.

Scout's vision had not returned, however, and, little though it was before he was ill, we soon realised just how much it had helped him. Scout's once-blurry world had now been plunged into total darkness, and we had to come up with a plan B to try and help him adjust to this change.

There was nothing else for it than to start again from scratch ... which is precisely what we did.

Beginning by walking Scout around the perimeter of the garden so that he could mind-map the area, we repeated the process in every room of our house. It took several attempts, and a few bumps and knocks but, true to form, Scout managed to re-establish his bearings over the next couple of days.

It saddened me greatly to think that Scout's already shadowy world had gotten even darker, but, at the end of the day, it was just another obstacle for us to overcome: I knew that we would deal with it together.

After several more days of total blindness, however, it became apparent that circumstances were changing yet again, which I originally thought was because Scout was simply adapting, as I noticed he seemed better able to negotiate his way around. But it seemed that Scout's condition was reverting to what it was before he became ill, and with it came the confidence and ebullience we had been missing.

I really wanted to understand what had happened to Scout's vision during his illness, and why a stomach upset had rendered him totally sightless. Scout's right eye was checked when he was poorly, and when he had recovered, and, in each case, there was very little response when a bright light was shone into his eye, and a complete lack of reaction to a menace response (the instinctive, protective blink reflex when something approaches the eye). Scout's vets had always found it difficult to determine exactly what he *could* see because of this complete lack of response, and the fact that his other senses tended to compensate for his lack of vision. Collectively, they agreed that there was *some vision*, be this an ability to differentiate between light and shade, or perhaps make out blurred images. Scout's left eye was totally sightless; this they knew for sure.

With no obvious physical changes to account for it, what had caused the sudden-onset blindness ... and why had Scout's vision now reverted to its original state?

The answer to this enigma is quite remarkable, and further evidence of Scout's determination and strength of character.

It would seem that Scout had to work hard to maximise what little sight he had by concentrating very intently on making the most of it and using it to the best of his ability. But a serious distraction, such as the debilitating tummy ache, meant he was unable to concentrate all his efforts on trying to see, because his reserves were redirected toward trying to deal with his illness. Once on the road to recovery, Scout began once more to turn his attention to maximising his limited sight. The theory does sound a little far-fetched, I confess, but really does seem to be the most likely explanation.

Several months after this episode, Scout suffered another bout of colicky tummy, which, although not as severe, had the same effect on his sight, the blindness staying with him for several days.

I have always thought of Scout as incomparable: a one-off. A once-in-a-lifetime dog, destined to test our knowledge and understanding of the animal within. Scout continues to astound and amaze us with his uniqueness and ability to make us smile.

18 Scout the comedian

Scout has had some moments of pure comedy gold.

Usually so careful never to leave anything out of place around the house, I had a momentary lapse one day, and left a low-sided plastic wash basket in the hallway. Paul and I were in the kitchen when we heard Scout charging about, his nails skittering on the wooden floor of the hallway. He collided with the basket, and – I'm not quite sure how – flipped into it, landing with sufficient momentum to propel him and the basket down the hall. We could scarcely believe our eyes as Scout and the basket whizzed past the doorway to the kitchen, Scout standing proudly at the helm, like the captain of a ship. Once the pair of them came to a halt, Scout calmly stepped out, shook himself, and nonchalantly wandered off.

My second indiscretion was to leave a bucket of soapy water unsupervised.

In my defence, I had checked on Scout's whereabouts before filling the bucket, and found him curled up on the sofa, apparently fast asleep. I was just about to carry the bucket out onto the front garden to wash my car when the telephone rang. Carefully setting down the bucket by the back door, I went to answer the call.

> **❝ ... We could scarcely believe our eyes as Scout and the basket whizzed past the doorway to the kitchen, Scout standing proudly at the helm ... ❞**

It can only have been a minute or two later that I heard a commotion. Ending the call I rushed outside to find the once-full bucket of sudsy water lying on its side in a frothy puddle. I called Scout's name, and he came hurtling over from the other side of the garden, sporting a foamy hat and beard. He looked very comical, but I did feel more than a twinge of guilt as I towel-dried his face and paws.

19 Spreading light

Several UK charities provide a much-needed service to local communities, and Pets as Therapy (PAT) is one of the largest organisations of its type. Registered volunteers and their temperamentally-assessed companion cats and dogs are invited to visit residents of hospitals, schools, and care homes, and provide invaluable companionship and love to everyone they meet.

Research has shown that these extremely worthwhile visits enhance the health and wellbeing of those who receive them, combating feelings of loneliness, depression, and anxiety, as well as helping with even more serious mental issues. The animals enrich the lives of those who spend time with them, and can also be real confidence boosters for children.

I had often thought that Scout would make the perfect PAT dog: with his affectionate nature, and ability to adjust the way he interacted with different people, he seemed to possess all of the necessary qualities.

And fate, it seemed, was about to step in once more to provide us with the opportunity to fulfil one last promise: that Scout would leave an indelible pawprint on the hearts of those in need of a friend.

In September 2016, we were invited by a lovely lady called Christine to attend the Malvern Autumn Show with Scout, Misty, and Sparky, to represent Whippets as part of the show's World of Animals exhibition, which showcased a plethora of different types of animal, including horses, pigs, sheep, cattle – and even the odd Alpaca, or two. A variety of dog breeds were invited to enable members of the public to meet and interact with as many as possible, and discover more about their traits and personalities. For those considering the acquisition of a specific breed of dog, this represented an ideal opportunity to find out more about their chosen breed.

As it happens, Christine is also a registered assessor for Pets as Therapy, and had agreed to carry out Scout's assessment for his suitability as a PAT dog. The criteria for a PAT dog is very exacting, and Scout would have to successfully pass a number of different behavioural assessments to be accepted for this important role.

We enjoyed a lovely morning at the Malvern showground. Situated just outside of the main showground, we were allocated our own stable with a little fenced-off enclosure. Christine was in the stable next to us with her own Whippets, and we were joined by Monty and his owner, Wendy. Monty is a huge and incredibly handsome Dogue De Bordeaux, who just happened to adore Whippets.

Scout, Sparky, and Misty spent the morning happily meeting and greeting members of the public. Although not a full pedigree Whippet,

Scout somehow managed to fulfil the role of honorary Whippet with his usual aplomb. He was fascinated by the activity around him, and all of the different aromas that continually flooded his senses. It goes without saying that he was enjoying all the fuss and attention that came his way.

Christine had arranged for Scout to have his assessment just after lunch, and this busy environment was ideal for assessing his walking skills (Scout had to demonstrate that he could walk in a calm and relaxed manner without pulling on the lead). Christine followed us as we skilfully threaded our way through the crowds, Scout keeping close to me at all times.

> *"... Scout did not appear overly-concerned when Christine dropped a large plastic bottle just behind him ... "*

I was feeling very confident about the next part of the assessment, in which Scout had to demonstrate his ability to accept being stroked and handled by strangers. This was no problem at all for Scout, of course, who always relished being touched and stroked.

Another small test was when Christine offered Scout a small treat, which he took without snatching or grabbing.

I was then asked to groom Scout, to again demonstrate his acceptance of close contact, and also to briefly hold him by his collar, as it may be necessary to restrain a PAT dog, should an emergency occur.

A PAT dog can potentially be exposed to loud and sudden noises, and needs to remain calm, should this happen. Scout did not appear overly-concerned when Christine dropped a large plastic bottle just behind him.

The final assessment was to see how Scout reacted to having to stand and wait whilst I engaged in a short conversation with Christine, demonstrating he could wait patiently without trying to move away. Scout responded to this situation by lying quietly at my feet.

The assessment was then repeated, but this time with Paul as Scout's handler, as he also wanted to be a volunteer. Once both assessments had been completed, we returned to our stable for a much-needed cup of tea, with Scout receiving a tasty chew as reward for his exemplary behaviour. All that was left to do to complete the application process was to submit the appropriate paperwork to Pets As Therapy ... and wait to hear if our application had been successful!

We ended our day at the Malvern show by taking part in the breed parade in one of the main arenas. Scout managed to sneak in and join the pedigree Hound group parade: he didn't care one bit whether or not he was a full pedigree, and cantered around the arena with the confident air of one who felt completely at home.

A few weeks after Scout was put through his paces at the Malvern show, we received the news we had been waiting for. Scout had passed his assessment, and was now a registered PAT dog! We could not have felt prouder when we received

Scout's registration certificate and identity badge. Who would ever have thought that our little dog, plucked from the streets as a young, vulnerable unwanted puppy, would go on to serve his community in such a positive way?

> **" ... Scout loved recieving this new attention, and his tail didn't stop wagging ... "**

Scout's services as a fully-fledged PAT dog were soon called upon. Jane, a close friend, is an events organiser at Saffron House, one of our local residential homes, and had been waiting patiently for Scout to become a PAT dog. Scout carried out his first visit to the home just before Christmas 2016. Wearing his best blue collar, complete with official identity badge, and sporting a festive green elf coat, Scout was introduced to the residents of the home.

I must confess I was a little nervous about the visit, and unsure what to expect when we arrived at Saffron House, but a warm welcome from Jane soon settled my nerves. After signing in, we were introduced to some of the members of staff. An expert meeter and greeter, Scout loved receiving this new attention, and his tail didn't stop wagging.

Once the initial greetings had been completed, Jane escorted us to the first of the home's two communal lounge areas, where many of the residents were watching the television. Jane announced our arrival, and, as we walked in, I saw the residents' faces light up as they spotted Scout, making his grand entrance.

The heart-warming sight of many of the residents lowering a hand, ready to stroke Scout as he visited each one of them in turn, will stay with me. Slowly and carefully, we made our way around the room, stopping at each resident to allow them to stroke Scout, who behaved like the perfect gentleman he is, standing completely still each time a hand was placed on the back of his neck, and moving away only when asked to do so.

The residents fell in love with Scout. Seeing the smiles on their faces, and hearing wonderful stories about the dogs they had owned and loved during their lifetimes, made me feel very humble indeed, and I did have to wipe away the occasional tear.

Once we had visited all of the residents in the lounge, Jane took us to visit others in their room. I wasn't really prepared for what Scout did here: I thought it was a fluke the first time, but he repeated the behaviour in every room we went into.

Jane knocked on the door before we entered each room, and confirmed that the resident was happy to meet Scout. Inside, I positioned Scout next to either the resident's bed or chair, and, as he had done in the lounge, he stood quietly and accepted a gentle stroke. The difference here was that, after standing for a few minutes, Scout lay down at the side of the bed or chair, emitted a loud sigh, and went to sleep, looking for all the world like he belonged there. It was almost as if he knew that, just by being present, he was providing gentle companionship in an unobtrusive way. Somehow,

Scout had worked out exactly what was needed of him, and delivered it with perfect timing, as always.

Coming to the end of our visit, Scout said hello to one or two residents in wheelchairs, positioning himself alongside the chairs in such a way that he could be touched and stroked. I could see he was enjoying this first visit immensely, and probably didn't want it to end.

Eventually, Jane took us back to the reception area, where she presented me with some tasty-looking homemade mince pies, and Scout received a Christmas card from the residents. That card is kept in a memory box we have made for Scout, together with all the cards, certificates, and letters of friendship that Scout has received during his life.

Our final promise to share the love that Scout gives so freely to all he meets was fulfilled in the best way possible.

20 Promises kept

As I come to the end of my musings about Scout, his life and his friends, my wish, dear reader, is that you will have taken our unique and wonderful little dog into your heart. Our journey may have been littered with obstacles and challenges, but courage and conviction have enabled us to overcome these.

❝ ... I reflect on Scout's life, and all that he has achieved, with an enormous sense of pride ... ❞

Scout is sleeping beside me as I write, his head resting against the side of my leg. Sparky is curled up at his side, and Misty is close by. I reflect on Scout's life, and all that he has achieved, with an enormous sense of pride. Little did we know that this little scrap of a dog would amaze and astound us in so many different ways. Scout has changed our lives for the better. He has taught us to follow our dreams, and make the impossible possible. Scout is our beacon of hope in what can sometimes be a dark world.

I watch Scout as his whiskers start to twitch, I think he is dreaming.

What do you see in your dreams, Scout? Do you see open fields of lush, green grass underneath a clear blue sky? Do you see golden, sandy beaches and tree-filled forests? Are your friends with you, Scout, helping to guide you? I would like to think so.

I don't know what new adventures fate has planned for us, though I do know that, whatever challenges come our way, we will face them together.

Since he arrived in our lives, all that we have done and continue to do have been for the love of our small dog, Scout, and the promises we made him.

Index

My dog is blind – but lives life to the full!

This invaluable book shows the owner of the newly-blind, partially-sighted or already blind dog that their loyal friend has lost none of her zest for life. With love and careful thought, you and your dog can get as much out of life as you always have, having fun and establishing an even closer bond. Includes case histories.

978-1-845842-91-8 £10.99*

Charlie – the dog who came in from the wild

The heart-warming true story of how one-eyed Charlie went from traumatised feral dog to joyful family member, bonding with the author, her daughter; making new human and canine friends, and eventually overcoming his fears to settle in to his new life.

978-1-845847-84-5 £10.99*

For more info on Hubble and Hattie books please visit www.hubbleandhattie.com; email info@hubbleandhattie.com *Prices subject to change